QUALITATIVE
MOTION
UNDERSTANDING

T0332513

QUALITATIVE MOTION UNDERSTANDING

Wilhelm BURGER
Johannes Kepler University
Linz, Austria

Bir BHANU
University of California
Riverside, California, USA

KLUWER ACADEMIC PUBLISHERS
Boston/London/Dordrecht

Distributors for North America:
Kluwer Academic Publishers
101 Philip Drive
Assinippi Park
Norwell, Massachusetts 02061 USA

Distributors for all other countries:
Kluwer Academic Publishers Group
Distribution Centre
Post Office Box 322
3300 AH Dordrecht, THE NETHERLANDS

Library of Congress Cataloging-in-Publication Data

Burger, Wilhelm.
    Qualitative motion planning / Wilhelm Burger, Bir Bhanu.
        p. cm. --(The Kluwer international series in engineering and
    computer science ; 184)
        Includes bibliographical references and index.
        ISBN 0-7923-9251-5 (alk. paper)
        1. Robots--Motion. 2. Computer vision. 3. Artificial Intelligence.
    I. Bhanu, Bir. II. Title. III. Series
    TJ211.4.B87   1992
    629.8'92--dc20   92-13870
                                                            CIP

Printed on acid-free paper.

Printed in the United States of America

CONTENTS

LIST OF FIGURES

Chapter 6

Chapter 7

Chapter 8

Appendix

PREFACE

Mobile robots operating in real-world, outdoor scenarios depend on dynamic scene understanding for detecting and avoiding obstacles, recognizing landmarks, acquiring models, and for detecting and tracking moving objects. Motion understanding has been an active research area for more than a decade to solve some of these problems. However, it still remains one of the more difficult and challenging areas of computer vision research.

Most of the previous work on motion analysis has used one of two techniques. The first technique employs numeric methods for the reconstruction of 3-D motion and scene structure from perspective 2-D image sequences. The structure and motion of a rigid object are computed simultaneously by solving systems of linear or nonlinear equations. This technique has been reported to be noise sensitive even when more than two frames are used. The second technique relies on the characteristic expansion patterns experienced by a moving observer. The basic idea is that, when a camera moves forward along a straight line in space, every point in the image seems to diverge from a single point, called the focus of expansion (FOE), and each image point's rate of expansion depends on the point's location in the field-of-view and the distance between the robot and the point.

This book describes a qualitative approach to dynamic scene and motion analysis, called DRIVE (Dynamic Reasoning from Integrated Visual Evidence). The DRIVE system addresses the problems of *(a)* estimating the robot's egomotion, *(b)* reconstructing the observed 3-D scene structure, and *(c)* evaluating the motion of individual objects from a sequence of monocular images. The approach is based on the FOE-concept but it takes a somewhat unconventional route. The DRIVE system uses a qualitative scene model and a fuzzy focus of expansion to estimate robot motion from visual cues, to detect and track moving objects, and to construct and maintain a global dynamic reference model.

The DRIVE approach consists of a two-stage process starting with given sets of displacement vectors for distinct image features in successive frames. First, the robot's egomotion (i.e., self-motion) is computed. To cope with the errors in feature correspondence, derotation and discretization, it uses a *Fuzzy FOE* concept, which defines an image region rather than a single point. In the second stage, DRIVE incrementally builds a *Qualitative Scene Model* (QSM) that is a camera centered 3-D model of the environment which describes the scene in qualitative terms, such as the relative distances of environmental entities and how these entities move in 3-D space. The qualitative scene model is declarative, describing the status and behavior of its elements and the relationships among them in coarse, qualitative terms. It does not try to derive a precise geometric description of the scene in terms of the 3-D structure and object motion. Using a mainly qualitative strategy of reasoning and modeling, our model allows for multiple simultaneous scene interpretations. Features that are believed to be part of the static environment are labeled and are used as references for computing the FOE.

The second stage also allows for the detection of moving objects in the scene. However, unlike traditional techniques for object tracking – like a multimodal tracking approach, where techniques such as centroid tracking, silhouette matching, correlation matching, feature matching, and Kalman filtering are synergistically combined and no 3-D information is used – DRIVE uses qualitative 3-D information to detect and track moving objects. Traditional tracking techniques may work satisfactorily in simple situations if the prediction is good. However, they do not make use of the sensor motion or the 3-D scene structure, and are unlikely to be effective in general scenes where there is no simple figure-ground relationship. It is expected that 3-D motion estimation and prediction will significantly help in tracking moving objects in cluttered natural scenes. We believe that qualitative approaches do have the potential to solve practical problems in dynamic scene and motion analysis. The work described here is an encouragement to proceed in that direction.

Chapter 1 provides the introduction to qualitative motion understanding, multi-level vision and motion analysis. Chapter 2 provides a framework for qualitative motion understanding. The main objectives of the approach are described, in particular the computation of camera motion, the detection of moving objects, and the estimation of stationary 3-D structure. Chapter 3 investigates the changes that occur in an image due to rotation and translation of the camera. It forms the basis for Chapter 4 where the decomposition of image motion into rotational and translational components is carried out. Chapter 4 develops two different approaches, *FOE-from-Rotation* and *Rotation-from-FOE*, for determining the 3-D translation and rotation of the camera. Chapter 5 proposes a "Fuzzy FOE" concept to overcome the difficult problem of finding the exact location of FOE in real image sequences. It contains a practical algorithm and presents results on real image sequences. Chapter 6 describes the processes for abstracting and interpreting image events. It discusses mechanisms for reasoning about 3-D scene structure and about 3-D motion. Chapter 7 presents the basic elements of the qualitative scene model, its dynamic evolution, strategies for conflict resolution, and the representation of multiple scene interpretations. Chapter 8 provides results on simulated data and real data obtained from a moving vehicle. Chapter 9 presents a summary of the work and points out several possible extensions. The appendices at the end of the book describe a geometric constraint method for estimating the camera motion and the absolute velocity of the vehicle.

The authors are grateful to Honeywell Systems and Research Center in Minneapolis, Minnesota, USA, where the technical work described in this book was performed. The work at Honeywell Inc. was supported by DARPA contract DACA76-86-C-0017 and monitored by U.S. Army Engineer Topographic Laboratories. The authors would like to thank DARPA and Bob Simpson for their support, enthusiasm, and encouragement. The authors would also like to thank Peter Symosek, John Kim, Hatem Nasr, Mark Gluch, Chris Meier, and Stephanie Schaffer for helpful suggestions during the development of the DRIVE system. Steve Axelson and John Kim helped in generating feature data used for experimentation in this work. John Anderson, Matthew Barth, Tom Payne and Subhodev Das provided useful comments. The book was developed while the authors were employed by Johannes Kepler University, Austria and the University of California, Riverside, USA. The authors are grateful to Prof. Franz Pichler at Johannes Kepler University, and Dean Susan Hackwood at the University of California for providing the facilities to prepare the manuscript.

1

INTRODUCTION

1.1 AIMS OF MOTION UNDERSTANDING

In recent years, *motion analysis* has established itself as an important discipline within the field of computer vision research. While the massive amount of data associated with the processing of real image sequences has long been a prohibitive factor, advances in computing technology have largely overcome this problem and made actual implementations feasible.

One of the key applications of machine vision is to facilitate the construction of machines which are capable of performing specific tasks without human guidance or interaction. Robots of the present generation are generally programmed "off-line" and operate without direct interaction with their environment, which limits their use to relatively simple, mostly repetitive tasks.

Why Motion?

Motion becomes a natural component of visual information processing as soon as moving objects are encountered in some form, e.g., while following a convoy, approaching other vehicles, or detecting moving threats. The presence of moving objects and their behavior must be known in order to provide appropriate reaction by an autonomous mobile robot. In addition, image motion provides important clues about the spatial layout of the environment and also about the actual movements of the robot. As part of the robot's control loop, visual motion feedback is essential for path stabilization, steering, and braking. Results from psychophysics [42, 58] show that humans also rely strongly on visual motion for motor control.

When the robot itself is moving, the entire camera image is changing continuously, even if the observed part of the environment is completely static. Therefore, the task of a mobile robot's vision system is the continuous interpretation of complex dynamic scenes. During the (monocular) imaging process, the direct perception of space is lost and any interpretation is solely based on a 2-D projection of the scene. However, any change in the 2-D image is always the result of a change in 3-D space, either induced by robot's motion or by moving objects. The objective of "Motion Understanding" is to find plausible interpretations for every change in the image.

1.2   AUTONOMOUS LAND VEHICLE NAVIGATION

The *Autonomous Land Vehicle* (ALV) project, sponsored by the Defense Advanced Research Projects Agency (DARPA), was intended to serve as a testbed for the ongoing research on mobile robots related to perception and planning. It provided the opportunity to integrate new implementations into an existing system and to demonstrate their performance in a real environment (Figure 1.1).

The ALV moves within a 3-D environment which contains stationary and moving objects at a broad range of distances. Visual information is provided by a regular TV camera which is mounted on top of the vehicle and whose focal length and height above the ground are known. The orientation of the camera relative to the ALV is fixed. Only coarse estimates of the vehicle's current position, orientation, and velocity are supposed to be given and no detailed map of the environment is available.

Visual information from the environment is an indispensable clue for the operation of the ALV in many regards. For example, even with the use of sophisticated inertial navigation systems, the accumulation of position errors requires periodic corrections. Mission tasks involving search and rescue, exploration and manipulation (e.g., refueling and munitions deployment) critically depend on visual information.

The primary goal of motion understanding in this specific task environment is to construct and maintain consistent and plausible interpretations of the time-varying images that are received by the moving camera. In particular,

Figure 1.1 The *Autonomous Land Vehicle* on a typical mission.

this involves answering questions like:

- What is moving in the scene and how does it move?

- How is the vehicle itself moving (i.e., the *ego-motion* of the vehicle)?

- What is the approximate 3-D structure of the scene?

Obviously, these three goals are in very close interaction. Any form of motion, whether ego-motion (camera motion) or actual object motion, must be measured against some stationary reference in the environment. Since the vehicle (and the camera) is moving, the projections of stationary objects in 3-D are generally not stationary on image the plane of the camera. In addition, the scene may contain any number of moving objects, such that it is not immediately clear what is moving in the environment and what can be trusted to be stationary. It is, therefore, one of the central issues in our approach to construct and maintain an internal *model* of that part of the environment which is *believed* to be stationary.

1.3   MULTI-LEVEL VISION AND MOTION ANALYSIS

Computer Vision is commonly structured into low-level and high-level processes. *Low-level* processes are purely 2-D, bottom-up and image-centered, such as feature extraction, feature matching or optical flow computation. *High-level* processes are 3-D and world-centered, attempting the semantic interpretation of the observed scene. Long-term observation and understanding of the behavior of objects form the basis for intelligent actions, e.g., navigation, route planning, and threat assessment.

In order to bridge the representational gap between the low and the high vision levels, processes operating at a *mediate* level have been defined (Figure 1.2). They are characterized by the transition from image-centered *features* to world-centered *objects*. The processes at the mediate level perform routine tasks that do not require special attention, but require some form of knowledge and short-term memory. *High-level* processes are activated whenever inconsistencies cannot be handled at the mediate level and closer attention is necessary.

Layers of Motion Analysis

As far as motion is concerned, each change in the 2-D image should be explained in terms of changes in the observed 3-D environment. Results should be made available to higher-level processes to allow long-term reasoning about the behavior of individual objects and their temporal interactions. It is the main task of mediate-level motion analysis to make sure that changes in the image are consistent with the current perception of the environment. Deviations from the expected behavior should result in an update of the internal model and reporting to the high-level processes, if necessary.

The processes active at the mediate level can be related to tasks that are performed unconsciously and automatically by the human visual system but require support from models, beliefs, and knowledge. For example, a person walking down a hall pays little attention to the behavior of the wall, because it is *believed* to be part of the stationary environment. However, a door may open at some point in time, signaled by a change in the image that cannot be accounted for in a completely stationary view of the environment. This is the point when attention, and thus higher levels of vision, come into the play.

Level	Representation	Tasks
High	World-Centered (3-D)	Semantic understanding; intelligent action; long-term observation; planning; attentative and knowledge-based, goal-directed.
Mediate	World/Viewer-Centered (2-D → 3-D)	Routine tasks; short-term observation; alerting; "stationary world maintenance;" non-attentative; shallow knowledge.
Low	Image-Centered (2-D)	Data driven, preprocessing; knowledge-free.

Figure 1.2 Taxonomy of vision processes. *Mediate-level* processes bridge the gap between image-centered, *low-level* processes and world-centered, *high-level* processes. The processes described in this book are mainly located at the two lower levels.

While the human is walking, however, the image on his retina keeps changing continuously, even though nothing in the environment changes, except the position and orientation of the eye. As a matter of fact, the image changes due to actual movement of the objects in the scene will often be less dramatic than the changes induced by the observer's motion at the same time. The conclusion is that 3-D perception and some form of "mental model" of the environment [30] plays an important role in the detection of 3-D object motion.

High-level and mediate-level processes are not only different with respect to the degree of abstraction but, as far as motion is concerned, also with respect to their *scope in time*. While the mediate level deals with short-term events, not extending beyond a certain time limit (in the range of seconds), the high level keeps a more global and long-term view of the events in the environment.

Bottom-Up vs. Top-Down

The information flow from the low-level processes to mediate-level processes is considered to be purely *bottom-up*. Interaction of mediate level processes with the high-level processes, however, is both *bottom-up and top-down*.

Bottom-up information is supplied by the mediate-level in the form of the current 3-D model of the environment, estimates for ego-motion, and indicators for object-motion. The high-level processes use this information for long-term navigation (planning), response, and expectations. In return, areas of interest might be identified (Focus of Attention) and "handed down" to the mediate level for more detailed analysis in a top-down fashion. As an example, the high level might be expecting the reappearance of an "interesting" object after an extended period of occlusion (e.g., a truck behind a hill). Consequently, processes at the mediate level will be advised to look out for motion in a certain area. The result of this observation is then reported back to the high level.

1.4   APPROACHES TO MOTION UNDERSTANDING

Conventional Approaches

Previous work in motion analysis has mainly concentrated on quantitative approaches for the reconstruction of motion and scene structure from image sequences – the review by Nagel [51] provides a comprehensive coverage. While most previous work on the reconstruction of camera motion assumes a completely stationary environment, here we allow the presence of moving objects. Since one cannot rely on a fixed camera setup to *detect* those moving objects, some kind of common reference is required to which the movement of the vehicle as well as the movements of objects in the scene can be related.

Extensive work has been done in determining relative motion and rigid 3-D structure from a set of image points and their displacements, usually involving one of the following two approaches:

- In the *first* approach, 3-D structure and motion are computed in one integral step by solving a system of linear or nonlinear equations [46, 69] from a minimum number of points on a rigid object. The method is reportedly sensitive to noise [27, 78]. Recent work has addressed the problem of recovering and refining 3-D structure from motion over extended periods of time, demonstrating that fairly robust results can be obtained [16, 18, 33, 65, 72]. However, these approaches require distinct views of the object (the environment), which are generally not available to a moving vehicle. In addition, it seems that the noise problem cannot be overcome by simply increasing the time of observation.

- The *second* approach [16, 36, 42, 43, 56, 59] makes use of the unique expansion pattern which is experienced by a moving observer. Arbitrary observer motion can be decomposed into translational and rotational components from the 2-D image without computing the structure of the scene. In the case of pure camera translation in a stationary environment, every point in the image seems to expand from one particular image location termed the *Focus of Expansion* (FOE). The closer a point is in 3-D, the more rapidly its image diverges away from the FOE. Thus for a stationary scene, the 3-D structure can be obtained directly from the expansion pattern. Certain forms of 3-D motion become apparent by local deviations from the expanding displacement field and can be therefore detected immediately.

The Qualitative Approach

The approach presented in this book is based on the FOE concept but takes a somewhat unconventional route. Like many other researchers in the field, we assume that individual image features can be tracked between successive images, resulting in a set of displacement vectors for each pair of images. However, in contrast to the classic technique for motion and scene structure, the computations of vehicle motion and scene structure are effectively decoupled. First, the *approximate* direction of vehicle heading is computed, resulting in what has been termed the "Fuzzy Focus of Expansion" (FFOE). This numerical computation is a 2-D process that employs no 3-D variables. Then, the apparent motion of image features with respect to the FFOE is encoded into *qualitative* (i.e., *symbolic*) descriptions that are fed into a rule-based reasoning system.

The blackboard system used for reasoning creates and manipulates a 3-D description of the scene, which is rather a *topological* than a precise geometrical representation. There are two interesting facts about this *Qualitative Scene Model* that is incrementally built over time. First, it allows efficient detection of static obstacles and moving objects in the scene. Second, it takes care of potential ambiguities by maintaining *multiple* scene interpretations within a single model. A truth maintenance technique is used to eliminate unfeasible scene interpretations.

1.5   OUTLINE OF THIS BOOK

In the following Chapter 2, we motivate and develop a framework for qualitative motion analysis. The main objectives of this approach are described, in particular the computation of camera motion, the detection of moving objects, and the estimation of stationary 3-D scene structure. The aspects of qualitative representation and reasoning are briefly summarized. Chapter 3 investigates the changes that occur in the image due to the camera's ego-motion, in particular the effects of camera translation and rotation and the concept of the *Focus of Expansion*. This forms the basis for Chapter 4, where we investigate the problems related to computing the camera's ego-motion from the given image motion.

In Chapter 4, we develop two different approaches towards computing vehicle translation and rotation. It turns out that finding the exact location of the FOE in real image sequences poses a very difficult problem. As a solution we propose a scheme involving a "Fuzzy" FOE, to be covered in Chapter 5, where we develop a practical algorithm and also present results from real image sequences.

In Chapter 6, we describe the transition from continuous image data to symbolic tokens, called *image events*. We show how these events can be used to reason about 3-D scene structure and motion, and how they contribute to the creation and maintenance of a *qualitative* model of the scene. The central aspects of this model, in particular the handling of ambiguities and the evolution of the model over time, are covered in Chapter 7.

A prototype system has been built to demonstrate the viability of this approach, and Chapter 8 presents two examples for the operation of that system on simulated and real image sequences. Finally, Chapter 9 provides a brief summary of this work and points out possible directions to future work.

In the Appendix of this book we have provided some auxiliary material that is related to but is not central to our approach. The first Appendix (A.1) describes an alternative method for computing the camera motion parameters by using a geometric constraint method, while the second Appendix (A.2) deals with the problem of estimating the absolute velocity of the vehicle.

2

FRAMEWORK FOR QUALITATIVE MOTION UNDERSTANDING

2.1 MOVING THROUGH A CHANGING ENVIRONMENT

The images that are acquired by a camera mounted on a moving vehicle are subject to continuous change. Even if the observed environment is completely static, the location of any environmental point projected onto the camera's image plane undergoes some movement as the vehicle travels along. In many practical applications, however, the environment cannot be considered completely static but may contain other moving objects, such as humans, birds, cars, machines, etc. Even the movements of trees on a windy day or the changing position of the sun would violate the static assumption.

Everything Moves

Clearly, the sequences seen by a viewer moving through such a changing environment may be arbitrarily complex. The aim of motion understanding, as pointed out earlier, is to explain any observed image motion in terms of the possible causes in the 3-D environment. At any point in time, the system should know whether a change in the image was caused by its own movement or by the movement of the corresponding entity out there in the real world. Figure 2.1 shows the traces of image features as they would typically be observed by a moving camera in a land vehicle application. Notice that all entities are stationary in this scene.

Image Plane

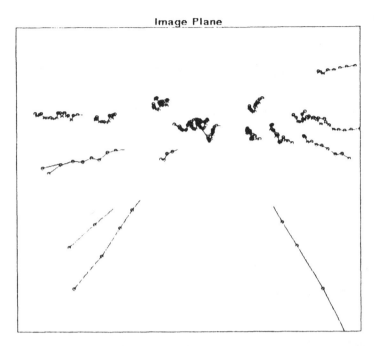

Figure 2.1 Traces of image features typically observed by a moving camera. While all entities in the corresponding scene are stationary, the detection of possible 3-D motion in such a sequence is not a trivial problem.

Put in other words, the problem is to decide what is actually moving in the scene and what is not. As we all know from personal experience, there is not always a clear answer to this question. For example, picture yourself on a train that is waiting at a station. When a train on a neighboring track starts to move, we may well have the illusion that our own train is pulling out of the station. Given no additional information, either of the two hypotheses could be the right one, i.e., either the own train or the other train is moving. In fact, *both* trains could be moving. However, as soon as we can find any *stationary* reference in the scene (e.g., a piece of a building), most possible scene interpretations will become inconsistent and would be discarded in favor of a single and consistent hypothesis.

Ambiguous Interpretations

Intuitively, it seems obvious how to distinguish between stationary and non-stationary things in the environment, at least in such a familiar setting as driving a vehicle. Formally, of course, one could treat the stationary part of the environment just like any other object moving relative to the observer, the observer himself being the only reference.

However, in addition to its intuitive appeal, we think that there are also practical advantages of treating the stationary world special. *First*, the vehicle must navigate from one point in the stationary world to another, irrespective of what moving objects it encounters. *Second*, any stationary point in the scene could actually be moving along a (possibly complicated) trajectory that makes it only *look* stationary. If we wanted to account for each and every such possibility, the number of concurrent hypotheses could be enormous. It, therefore, makes sense to make the (heuristic) default assumption that things are *stationary* unless there is contradicting evidence.

Handling the inherent ambiguities in high-level motion analysis plays a central role in the framework being described below. There are situations, of course, where motion analysis is simply not powerful enough to produce unique scene interpretations. However, many ambiguous situations can be eliminated by simultaneous reasoning about hypothesized 3-D motion and stationary 3-D scene structure. The main elements of this approach are outlined in the following.

2.2 THE "DRIVE" APPROACH

The work described here is in fact part of a larger project at Honeywell Systems and Research Center (SRC) dealing with various aspects of dynamic scene analysis. The idea was to construct several processing modules for the different tasks (like landmark recognition, terrain interpretation, object recognition, motion analysis and tracking) that would work cooperatively for navigation in outdoor scenarios [12]. In this book we describe the motion analysis component of that project, termed the DRIVE[1] module [7, 8].

The DRIVE module operates frame-by-frame and is internally structured into three stages (Figure 2.2):

---

[1] DRIVE stands for *D*ynamic *R*easoning from *I*ntegrated *V*isual *E*vidence.

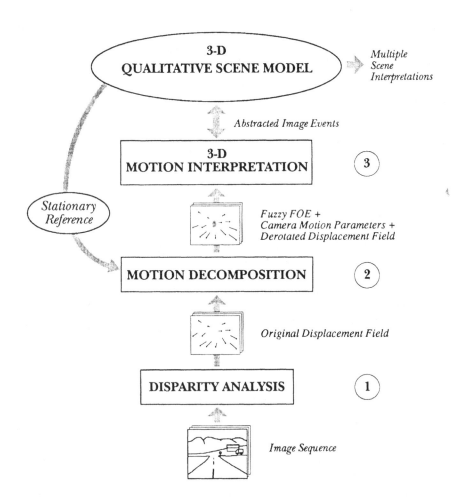

Figure 2.2 Main elements of the "DRIVE" approach. At the lowest level (*Disparity Analysis*), salient image points are selected and tracked between successive frames, thus producing a set of displacement vectors for each pair of frames. Subsequently (*Motion Decomposition*), the original displacement field is decomposed to compute the camera's motion parameters and to *derotate* the displacement field. The translational (derotated) displacement field is used by the *3-D Motion Interpretation* stage to create and update a 3-D description of the scene, called the *Qualitative Scene Model*, which generally maintains multiple scene interpretations. The model also supplies a set of stationary reference entities for computing the camera motion.

(1) *Disparity Analysis:* Distinct features are selected from the image by a low-level process and tracked from frame to frame, thus producing a field of displacement vectors for each pair of frames.

(2) *Motion Decomposition:* The displacement field obtained in stage *(1)* is *derotated* by removing the image effects caused by vehicle rotation. This reveals the location of the *Fuzzy Focus of Expansion*, which in turn provides the direction of vehicle translation. A set of entities believed to be stationary in the scene (supplied by the scene model) is used as the reference for the camera motion. From the displacement vectors of selected features on the ground the absolute velocity of the vehicle is determined. This allows the computation of all vehicle motion parameters in absolute terms.

(3) *3-D Motion Interpretation:* The derotated displacement field is examined for salient static and dynamic relationships between features. These "abstracted image events" are used to make inferences about the 3-D scene structure, individual object motion, and related information which is gathered in the *Qualitative Scene Model.* The scene model is continuously updated as the vehicle moves through its environment. It generally maintains multiple scene interpretations to cope with the possible ambiguities that we have mentioned earlier.

In the remaining parts of this chapter, we briefly sketch some of the general problems and motivations that we found relevant for developing the individual stages of the DRIVE approach.

2.3   LOW-LEVEL MOTION

The problem of estimating the 2-D motion in an image sequence has been a field of active research for many years [32, 67]. Two major techniques have been developed, which we refer to as the *Flow* and the *Displacement* methods.

The Flow Method

The *Flow Method* uses spatial and temporal variations of image intensity to estimate the *instantaneous velocity* at each pixel location, often called the *opti-*

cal flow field [34, 42, 55, 63]. In general, the method depends upon sufficient object texture, continuous motion, and small displacement between
subsequent frames. Unfortunately, for very small displacement in two perspective views the problem of estimating the 3-D structure of the scene
becomes ill-conditioned [51]. Because the magnitude of flow can only be
determined in the direction of the 2-D gradient (perpendicular to the tan-
gent of a boundary), the overall flow field cannot be computed using only
local information. Global smoothing of the flow field has been proposed,
which, in turn, creates problems at flow discontinuities, such as object
boundaries or transparent surfaces.

The Displacement Method

The *Displacement Method* [3, 6, 10, 27 39, 50] uses the parts of the image
where discontinuities in brightness or motion occur (which are a source of
problems in the flow method). Significant (or "interesting") features, such
as line segments, distinct dark or bright spots, or corners are selected in
consecutive frames and matched, thus producing a field of *displacement vectors* between corresponding features.

The displacement method employs a formulation of 2-D motion as discrete
transitions between successive image locations, as compared to the velocity-
based formulation which characterizes the flow method. With the displacement method, there is usually no problem related to motion discontinuities,
as with the flow scheme. In fact, interesting image features used for measuring the displacement are frequently located at exactly the motion boundaries, e.g., the object contours. However, two major problems are associated
with the displacement method: the problem of feature selection and the
corrspondence problem.

Feature Selection

The *first* problem related to the displacement method is selecting and
locating features reliably in successive frames, especially when the images are
noisy. Individual features are commonly extracted by applying local window
operations. Moravec's "Interest Operator" [48] for detecting salient points
is a classic example. In the given application, where the camera moves close
to the ground, the reliable extraction of interest points is not a simple task.
A point which was clearly distinguishable at one time may become blurred
when the camera gets closer to it, or even disintegrate into several tokens.

Features of this kind are very difficult to track. This problem suggests some form of *range-dependent* feature-extraction scheme, possibly by using some coarse assumptions about the spatial layout of the scene. For example, one could apply larger operator sizes in the lower parts of the image.

The Correspondence Problem

The *second* problem related to the displacement method is finding reliable matches between the set of interesting features extracted from one frame and the set of features found in the subsequent frame. This is commonly called the *Correspondence Problem* which is a difficult task in itself and has been the subject of intense research [6, 64, 70]. There is evidence that the problem is solvable and algorithms now exist that perform fully automatic feature selection and tracking with sufficient reliability. A successful algorithm was developed as part of this project. Unfortunately, this algorithm was not available for the experiments shown in the subsequent parts of this book, where features were selected and tracked manually. For experiments using automatic feature selection and tracking, the reader is referred to [12, 38].

In summary, we consider the *Displacement Method* to be more promising than the *Flow Method* for our purpose. Not only is the information contained in discrete displacements fields of more immediate use, but the problem of reliably computing the *optical flow* appears to be just as hard as extracting and matching distinct features [73].

2.4 CAMERA MOTION AND SCENE STRUCTURE

The foremost questions to be answered by a visual navigation system are related to *(a)* the spatial layout of the environment, and *(b)* the motion of the vehicle with respect to this environment. To control and maneuver the vehicle requires knowing its motion parameters at any point in time. The vision system should be an integral part of the vehicle control loop. Similarly, knowledge about the 3-D layout of the scene is essential for obstacle avoidance, navigation, and route planning.

While direct information about the 3-D distance of objects is not available in single, monocular images, a moving observer can get a spatial impression by an effect called *motion stereo*. When the observer performs translatory motion in 3-D space, different views of an object are obtained, thus allowing inferences about 3-D shapes. The computational approaches to exploit this effect fall into one of two categories, which we want to call the *Reconstruction Method* and the *FOE-Method*, respectively.

The Reconstruction Method

The *Reconstruction Method* assumes that the environment moves as a single rigid object relative to a stationary camera. This is, of course, equivalent to the notion of a stationary environment being viewed by a moving camera. The assumption of rigidness is a central one. The problem is to compute a 3-D configuration that is rigid and consistent with the observed sequence of 2-D projections. This is usually formulated as a minimization problem over a system of linear or non-linear constraints, mostly with a least-squares error function. The solution comprises the 3-D coordinates of each scene point plus the parameters of the camera motion. The simultaneous computation of camera motion and 3-D scene structure is characteristic of the Reconstruction Method.

Practical experiments [27, 78] show that the results obtained by these methods are often sensitive to noise in the image, which always exists in reality due to spatial sampling, camera distortion, blurring, etc. It has been shown that improvements are possible and that at least error estimates can be obtained [28, 77]. Various relationships have been established between the number of feature correspondences and the number of views needed to obtain unique 3-D solutions [44, 69, 71].

Multi-Frame Motion

Recent work has addressed the problem of recovering and refining 3-D structure from motion over multiple frames and has demonstrated that fairly robust results can be obtained [16, 33, 72]. However, these methods require large amounts of computation, and convergence is slow. Good performance usually requires that many disparate views of the object (i.e., the environment) be available, which is generally not the case in our domain. In addition, the noise problem persists even in multi-frame approaches.

More recently, various approaches using Kalman filtering have been developed for range computation from motion stereo [46, 66]. These approaches are computationally expensive and require accurate camera motion parameters.

The FOE-Method

The *FOE Method*, the basis of the DRIVE approach, makes use of the unique "retinal" expansion pattern that is experienced by a moving observer undergoing pure translation [16, 17, 36, 42, 43, 45, 56, 59].

Pure Camera Translation

Under pure translation, the direction of heading in 3-D is found at the *focus of expansion* (FOE), which is the image location where all displacement vectors intersect, i.e., image features corresponding to the stationary points in the scene seem to "expand" radially from the FOE under forward translation. This property of the displacement field makes it possible to compute the camera motion without computing the 3-D scene structure at the same time. Consequently, the computation of camera motion and scene structure are effectively decoupled in the FOE approach.

If the exact location of the FOE in the image is known, the relative depth of every (stationary) image point can be computed by taking the ratio of expansion vs. the distance from the FOE. A depth map and thus the structure of the scene is obtained without the need to solve large systems of equations. The views of the scene need not be very distinct in this approach and there seems to be evidence from psychophysics that the human visual system may employ similar techniques [54, 58].

Arbitrary Camera Motion

Of course, we cannot assume that the vehicle moves along a straight line in 3-D space. In addition to translating in an unknown direction, the vehicle also rotates about its three axes[1]. Due to the design of the vehicle, certain assumptions can be made about its motion. First, the direction of travel is quite restricted, i.e., it is unlikely to take off the ground or submerge into it. It can only move forward or backward, i.e., the direction of heading is normally within the camera's field of view. Second, the vehicle has considerable

---

[1] *Roll, pitch,* and *yaw.*

inertia such that its orientation cannot change rapidly. In addition, the rotations about the longitudinal axis Z are, as it turns out, small enough to be ignored entirely. Since the camera is rigidly mounted on the vehicle, *vehicle motion* and *camera motion* are identical – we use these two terms interchangeably.

The main problem associated with the FOE-method is to locate the FOE when the camera does *not* undergo pure translation but rotates as well. The effects of vehicle translation and rotation combine in the observed displacement field, such that the motion components cannot be determined from the displacement field directly. However, the displacement field caused by the vehicle's ego-motion can be decomposed into its rotational and translational components exclusively in the 2-D image, without any 3-D information [19, 56, 57, 60].

Unfortunately, finding the *exact* location of the FOE is generally impossible under real conditions. This leads us to formulate the concept of a "Fuzzy FOE", which accounts for the effects of noise and errors in the given displacement field by searching only for the approximate direction of camera heading. Of course, given only the *approximate* direction of camera heading, we can only compute the *approximate* 3-D distances of stationary scene entities.

Non-Stationary Scenes

In the past, the FOE-method has been studied almost exclusively with stationary scenes that contain no other moving objects [2, 5, 37, 41, 56, 61]. Adiv [1] has shown that the movements of other objects can be accounted for by computing multiple FOEs, one for each individual object. However, this method requires that at least a few features can be tracked on each object, which is often not possible due to the small image size of distant objects.

2.5 DETECTING 3-D MOTION

For intelligent action in the presence of potential threats and targets, or navigation in a traffic environment, information about 3-D motion in the scene is indispensable. Moving objects must be detected and isolated from the stationary environment and, if possible, their current motion should be estimated to track them and to create expectations about their future behavior.

Since the camera itself is moving, the stationary part of the environment cannot be assumed to be registered in subsequent images, as in the case of a stationary sensor. Simple frame-differencing techniques to detect and isolate moving objects do not work in this case because image changes due to sensor motion would generate too many false alarms. More sophisticated image-based techniques, which apply 2-D transformations (warping) to the image to compensate background motion, work well only when objects are moving in front of a relatively flat background, such as in some air-to-ground applications. To detect actual object motion in the complex scenario of a mobile robot, the 3-D structure of the observed environment together with the vehicle's ego-motion must be taken into account.

In our approach, 3-D motion is detected in one of two ways:

- *First*, some forms of motion are concluded directly from the 2-D displacement vectors without any knowledge about the underlying 3-D structure.

- *Second*, motion is detected by discovering inconsistencies between the current state of the internal 3-D scene model and the changes actually observed in the image.

2.6 QUALITATIVE MODELING AND REASONING

The choice of a suitable scheme for representing the perceived state of the scene is a crucial one for dynamic scene understanding. In our DRIVE approach, the environment is modeled as a 3-D, time-varying configuration of rigid objects whose structures, relative positions, and motions are estimated from visual information.

Stationary or Mobile

The stationary part of the world is represented by a subset of the objects, which form a rigid configuration in 3-D space. This definition, however, is not sufficient to identify the stationary world *a priori*, because more than one rigid subset of world objects may be observed. For example, looking at the back of a truck which travels closely in front of the observer could give the impression of a large object (i.e., the truck in front), whose image position is more or less constant. This object is rigid and does not move with respect to the observer, but it is not part of the stationary world.

Quantitative Modeling

Most previous work attempted to obtain the 3-D scene structure from motion in the form of a quantitative, numerical description of the spatial layout of the environment relative to the camera. The problems related with this approach are well-known and applications using real imagery have been rare. Numerical schemes for computing 3-D structure and camera motion (i.e., the *Reconstruction Method*) produce a single solution that is optimal in some sense. There seems to be no direct way to make the final result reflect the *uncertainty* contained in the input data. Furthermore, the assumption of rigidness in 3-D cannot be guaranteed to hold. When scene entities are assumed to form a rigid configuration in space, but are actually moving relative to each other, this may still produce a numerically acceptable interpretation. The interesting question is how this solution reacts when moving features and stationary features are inadvertently grouped. In the best case, this would cause a relatively high error for the optimal solution such that the solution could be rejected. However, there would be no explicit information that tells about which of the entities are actually in motion.

Qualitative Motion

We argue that a precise numerical description of the 3-D environment is not really necessary for efficient reasoning. Alternatively, the idea to use *qualitative* descriptions of real-world phenomena has raised considerable interest in the areas of mathematics, physics, and artificial intelligence [40]. Its potential significance to the field of computer vision has been addressed only recently [20, 68, 73]. The main argument is that computationally expensive (and often error–prone) techniques, as they are abundant in machine vision, can sometimes be replaced by emphasizing qualitative effects and using less precise representations, without sacrificing the relevance of the results.

Consequently, the central component of the DRIVE-approach is a *Qualitative Scene Model* (QSM) that is built incrementally and can be considered as the "mind" of the motion understanding system (see top of Figure 2.2). This model is a 3-D, camera-centered representation of the scene which describes the observed environment in simple qualitative terms. The set of entities in the QSM is conceptually split into two parts, the *stationary world* and a set of independently moving entities. Construction of the QSM over time is accomplished by a reasoning process which draws conclusions from significant events in the image.

Additionally, the state of the QSM is a not a single interpretation but a *set* of interpretations which are all pursued simultaneously. This provides a very flexible mechanism for handling the inherent ambiguities encountered in image understanding. Each interpretation is a collection of hypotheses, called *partial* interpretations, which cover overlapping subsets of the entities in the model. As the vehicle travels through the environment, the model is continuously updated and revised by adding or deleting hypotheses. Chapter 7 describes in more detail the structure and operation of this qualitative modeling scheme, which can be considered the core of the DRIVE approach.

3

EFFECTS OF CAMERA MOTION

Following the brief outline of the "DRIVE" approach in Chapter 2, its individual components (Figure 2.2) are now described in more detail. The geometry of the imaging process for a moving camera is covered in the first section, followed by a discussion of the image effects caused by individual camera rotation and translation. This provides the background for the motion decomposition algorithms that will be developed in Chapters 4 and 5.

3.1 VIEWING GEOMETRY

It is well-known that any rigid motion of an object in space between two points in time can be decomposed into a combination of translation and rotation. While many researchers have used a velocity-based formulation of the problem [1, 56, 59], the following treatment views motion in discrete time steps. Given the world coordinate system $(X\, Y\, Z)$ shown in Figure 3.1, a translation $\mathbf{T} = (U\, V\, W)^T$ applied to a 3-D point $\mathbf{X} = (X\, Y\, Z)^T$ is accomplished through vector addition:

$$\mathbf{X}' = \mathbf{T} + \mathbf{X} = \begin{bmatrix} X' \\ Y' \\ Z' \end{bmatrix} = \begin{bmatrix} U \\ V \\ W \end{bmatrix} + \begin{bmatrix} X \\ Y \\ Z \end{bmatrix} \tag{3.1}$$

A 3-D rotation $\mathbf{R}$ about an arbitrary axis through the origin of the coordinate system can be described by successive rotations $\mathbf{R}_\phi\, \mathbf{R}_\theta\, \mathbf{R}_\psi$ about its X-, Y-, and Z-axes, respectively. Thus

$$X' = R\,X = R_\phi\,R_\theta\,R_\psi\,X\,,\qquad(3.2)$$

where

$$R_\phi = \begin{bmatrix} 1 & 0 & 0 \\ 0 & \cos\phi & -\sin\phi \\ 0 & \sin\phi & \cos\phi \end{bmatrix}\quad\text{is a rotation about the } X\text{-axis,}\qquad(3.3a)$$

$$R_\theta = \begin{bmatrix} \cos\theta & 0 & \sin\theta \\ 0 & 1 & 0 \\ -\sin\theta & 0 & \cos\theta \end{bmatrix}\quad\text{is a rotation about the } Y\text{-axis,}\qquad(3.3b)$$

$$R_\psi = \begin{bmatrix} \cos\psi & \sin\psi & 0 \\ -\sin\psi & \cos\psi & 0 \\ 0 & 0 & 1 \end{bmatrix}\quad\text{is a rotation about the } Z\text{-axis.}\qquad(3.3c)$$

A general rigid motion in space consisting of translation and rotation is therefore described by the 3-D transformation M:

$$X \rightarrow X' = M(X) = R_\phi\,R_\theta\,R_\psi\,(T+X)\qquad(3.4)$$

Its six degrees of freedom are U, V, W, ϕ, θ, and ψ. This decomposition is not unique because the translation could be applied after the rotation instead. Also, since the multiplication of the rotation matrices is not commutative, a different order of rotations would result in different amounts of rotation for each axis. For a fixed order of application, however, this motion decomposition *is* unique.

To model the movements of the vehicle, the camera is considered as being stationary and the environment as being moving as a single rigid object relative to the camera. The origin of the coordinate system is located in the lens center of the camera (Figure 3.1). The given task is to reconstruct the vehicle's ego-motion in 3-D space from the motion information available in the 2-D image. It is therefore necessary to know the effects of different kinds of vehicle motion upon the observed image.

Under perspective imaging, a point $X = (X\ Y\ Z)^T$ in 3-D space is projected

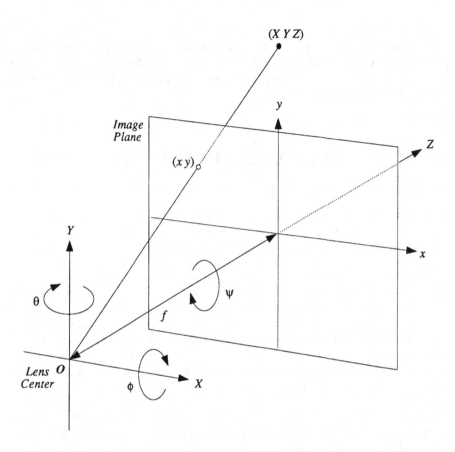

Figure 3.1 Viewing geometry and camera model. The origin of the coordinate system *O* is located at the lens center. The focal length *f* is the distance between the lens center and the image plane.

onto a location in the image plane $\mathbf{x} = (x\ y)^T$, with

$$x = f\frac{X}{Z} \qquad\qquad y = f\frac{Y}{Z} \qquad\qquad (3.5)$$

where *f* is the focal length of the camera (see Figure 3.1).

For the following discussion, a few definitions will be useful. The set of observed image points in a frame *k* is called an *image point set* (or *image* for short), denoted by $\mathbf{I}_k = \{\mathbf{x}_1, \mathbf{x}_2, \ldots, \mathbf{x}_N\}$, where $\mathbf{x}_i \neq \mathbf{x}_j$.

A *displacement vector* is a tuple $\mathbf{d}_i = (\mathbf{x}, \mathbf{x}') \in \mathbf{I}_k \times \mathbf{I}_{k+1}$ of corresponding image points in successive frames. Finally a *displacement vector field* $\mathbf{D}_k$ is a set of displacement vectors or, to be more precise, a 1:1 mapping between successive image point sets $\mathbf{I}_k$ and $\mathbf{I}_{k+1}$.

3.2 EFFECTS OF CAMERA ROTATION

When the camera is rotated around its lens center, the acquired image changes, but no additional, entirely new views of the environment are obtained. Ignoring the effects at image boundaries and effects due to discretization of the image, camera rotations merely map the image into itself.

The effects caused by pure camera rotation about one of the coordinate axes are intuitively easy to understand. For example, if the camera is rotated about the Z-axis (i.e., the optical axis), points in the image move along circles centered at the image location $\mathbf{x}_c = (0\ 0)$. In practice, however, we can assume that the amount of vehicle rotation about the Z-axis is small. Therefore, in this work, we ignore rotation about the Z-axis and consider only the more significant camera rotations about the X- and Y-axes.

Rotating the vehicle about the Y-axis by an angle $-\theta$ and about the X-axis by an angle $-\phi$ moves each 3-D point $\mathbf{X}$ to a new location $\mathbf{X}'$ in the camera-centered coordinate system:

$$\mathbf{X} \rightarrow \mathbf{X}' = \mathbf{R}_\phi\, \mathbf{R}_\theta\, \mathbf{X} = \begin{bmatrix} \cos\theta & 0 & \sin\theta \\ \sin\phi\,\sin\theta & \cos\phi & -\sin\theta\,\cos\theta \\ -\cos\phi\,\sin\theta & \sin\phi & \cos\phi\,\cos\theta \end{bmatrix} \cdot \begin{bmatrix} X \\ Y \\ Z \end{bmatrix} \tag{3.6}$$

Consequently $\mathbf{x}$, the projected image point of $\mathbf{X}$, moves to $\mathbf{x}' = (x'\ y')^T$:

$$\begin{bmatrix} x' \\ y' \end{bmatrix} = \frac{f}{-X\cos\phi\sin\theta + Y\sin\phi + Z\cos\phi\cos\theta} \begin{bmatrix} X\cos\theta + Z\sin\theta \\ X\sin\phi\sin\theta + Y\cos\phi - Z\sin\theta\cos\theta \end{bmatrix}$$

$$\tag{3.7}$$

By inverting the perspective transformation (3.5), we get

$$X = \frac{1}{f} Z x \qquad\qquad Y = \frac{1}{f} Z y \qquad\qquad (3.8)$$

for some image point $\mathbf{x} = (x\, y)$ and a given focal length f. Using (3.7) and (3.8), the 2-D movement of an image point $\mathbf{x} \to \mathbf{x}'$ caused by vertical and horizontal camera rotations (by angles ϕ and θ) can be expressed as a *rotational mapping* $\rho_{\phi\theta}$,

$$\mathbf{x} \to \mathbf{x}' = \rho_{\phi\theta}(\mathbf{x}) \qquad\qquad (3.9)$$

where

$$\begin{bmatrix} x' \\ y' \end{bmatrix} = \rho_{\phi\theta} \begin{bmatrix} x \\ y \end{bmatrix} \qquad\qquad (3.10)$$

$$= \frac{f}{-x\cos\phi\sin\theta + y\sin\phi + f\cos\phi\cos\theta} \begin{bmatrix} x\cos\theta + f\sin\theta \\ x\sin\phi\sin\theta + y\cos\phi - f\sin\theta\cos\theta \end{bmatrix}$$

Notice that the only parameters of this transformation are the original image position $(x\, y)$, the angles of rotation θ and ϕ, and the camera's focal length f. This means that the image effects of pure camera rotations can be computed without knowing the 3-D distance of the observed points from the image plane. It also illustrates the fact that camera rotations alone do not convey any information about the 3-D structure of the scene. However, $\rho_{\phi\theta}$ is a non-linear mapping in the image plane, while the corresponding 3-D rotations $\mathbf{R}_\theta$ and $\mathbf{R}_\phi$ are linear transformations performed by matrix-vector multiplications.

The transformation in (3.10) has another interesting property. Moving an image point along a diagonal that passes through the center of the image at 45° by only rotating the camera results in different amounts of rotation about the X- and the Y-axis. This is a consequence of applying the camera rotations $\mathbf{R}_\theta$ and $\mathbf{R}_\phi$ sequentially. The first rotation $\mathbf{R}_\theta$ (about the Y-axis) changes the orientation of the camera's X-axis in 3-D space before $\mathbf{R}_\phi$ is applied. This also explains why (3.7) is not symmetric with respect to θ and ϕ.

3.3 COMPUTING THE CAMERA ROTATION ANGLES

To obtain the angles of camera rotations from a pair of observations, we need to solve the inverse problem. Given are two image points x_0 and x_1, which are the projections of the same 3-D point at time t_0 and time t_1, respectively. The problem is to determine the angles of rotation θ and ϕ which, when applied to the camera between time t_0 and time t_1, would move image point x_0 onto x_1 (assuming that no camera translation has occurred).

When horizontal rotation R_θ (about the Y-axis) and vertical rotation R_ϕ (about the X-axis) are applied to the camera separately, points in the image move along hyperbolic paths [56], as shown in Figure 3.2. For example, rotating the camera about the Y-axis moves a given image point $x_0 = (x_0\, y_0)$ along a path $(x\, y)$, described by the equation

$$y^2 = y_0^2 \frac{f^2 + x^2}{f^2 + x_0^2} \tag{3.11}$$

Similarly, rotating the camera about the X-axis moves an image point $x_1 = (x_1\, y_1)$ along a path $(x\, y)$ described by

$$x^2 = x_1^2 \frac{f^2 + y^2}{f^2 + y_1^2} \tag{3.12}$$

Figure 3.3 illustrates the image effects of a composite 3-D camera rotation consisting of R_θ followed by R_ϕ. The horizontal rotation R_θ moves the original image point x_0 to the intermediate location x_c. The subsequent vertical rotation R_ϕ moves x_c to the final image location x_1. The image point x_c is located at the intersection of a "horizontal" hyperbola passing through x_0, described by (3.11), and a "vertical" hyperbola passing through x_1, described by (3.12). The coordinates of the intersection point x_c are

$$\begin{bmatrix} x_c \\ y_c \end{bmatrix} = \frac{f}{\sqrt{(f^2 + x_0^2)(f^2 + y_1^2) - x_1^2 y_0^2}} \begin{bmatrix} x_1 \sqrt{f^2 + x_0^2 + y_0^2} \\ y_0 \sqrt{f^2 + x_1^2 + y_1^2} \end{bmatrix} \tag{3.13}$$

(a)

Figure 3.2 Effects of pure camera rotation. *(a)* Pure horizontal rotation (about the Y-axis). *(b)* Pure vertical rotation (about the X-axis). When the camera is rotated about an axis parallel to the image plane, image points move along hyperbolic paths regardless of their location in 3-D space.

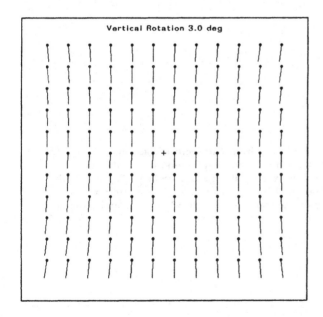

(b)

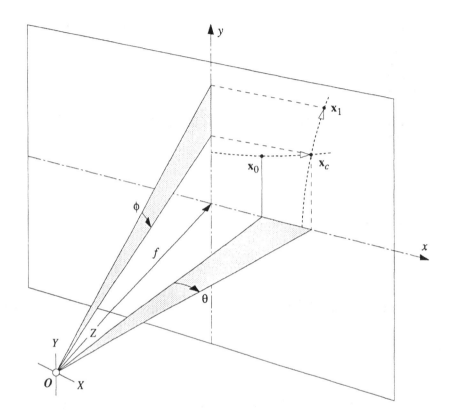

Figure 3.3 Successive application of horizontal and vertical rotation. An image feature is originally located at $\mathbf{x}_0$. Under horizontal rotation (about the Y-axis by an angle θ), the feature moves from $\mathbf{x}_0$ to $\mathbf{x}_c$ along a hyperbolic path. Under subsequent vertical rotation (about the X-axis by an angle ϕ), it moves from $\mathbf{x}_c$ to $\mathbf{x}_1$ The two rotation angles θ and ϕ can be computed directly when $\mathbf{x}_0$, $\mathbf{x}_1$, and f (the focal length of the camera) are known.

From this, the angles of camera horizontal and vertical rotation θ and ϕ for the observed image point motion $x_0 \rightarrow x_1$ (see Figure 3.3) are obtained directly as

$$\theta = \tan^{-1}\frac{x_c}{f} - \tan^{-1}\frac{x_0}{f} \qquad\qquad \phi = \tan^{-1}\frac{y_c}{f} - \tan^{-1}\frac{y_1}{f} \qquad (3.14)$$

3.4 EFFECTS OF CAMERA TRANSLATION

When the vehicle undergoes pure translation $\mathbf{T} = (U\ V\ W)^T$ between time t and time t', every point in the 3-D environment moves relative to the camera by the vector $-\mathbf{T}$. Since every stationary point is affected the same translation relative to the camera, these points can be thought of moving along imaginary parallel lines in 3-D space. One of the fundamental results from perspective geometry [25] is that the images of parallel lines pass through a single point in the image plane, called a *vanishing point*. Therefore, when the camera moves forward along a straight line in 3-D space, all (stationary) image points seem to diverge from a single vanishing point, called the *Focus of Expansion* (FOE). Alternatively, image points seem to converge towards a vanishing point, called the *Focus of Contraction* (FOC), when the camera moves backwards. Each displacement vector passes through the FOE (or FOC) creating the typical radial pattern that is illustrated in Figure 3.4.

As can be seen in Figure 3.4, the straight line passing through the lens center of the camera (O) and the FOE (F) is also parallel to the 3-D displacement vectors. Therefore, the 3-D vector $\overrightarrow{OF}$ (from the lens center to the FOE) points in the *direction* of camera translation $\mathbf{T}$ in 3-D space but does not supply the *length* of $\mathbf{T}$. The actual translation vector $\mathbf{T}$ is a multiple of the vector $\overrightarrow{OF}$:

$$\mathbf{T} = \lambda\ \overrightarrow{OF} = \lambda \begin{bmatrix} x_f \\ y_f \\ f \end{bmatrix} \qquad (3.15)$$

(for some scale factor $\lambda \in \mathbb{R}$). In velocity-based models of 3-D motion, the *focus of expansion* has commonly been interpreted as the *direction of instantaneous heading*, i.e., the direction of vehicle translation during an infinitely

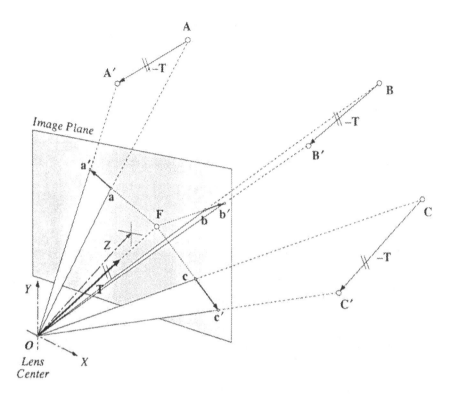

Figure 3.4 Location of the focus of expansion (FOE). With pure vehicle translation, points in the environment (**A**, **B**, **C**) move along 3-D vectors parallel to the vector pointing from the lens center (**O**) to the FOE (**F**) in the camera plane. These vectors form parallel lines in space which have a common vanishing point (the FOE) in the perspective image.

short period in time. When images are given as successive "snapshots" with significant camera motion between each frame, a discrete model seems more appropriate that treats the FOE as the direction of *accumulated* vehicle translation over a certain *period* of time.

The effects of camera translation **T** can be expressed as a mapping τ from a set of image points $\mathbf{I} = \{\mathbf{x}_1, \mathbf{x}_2, \ldots, \mathbf{x}_N\}$ onto a corresponding vector of image points $\mathbf{I}' = \{\mathbf{x}_1', \mathbf{x}_2', \ldots, \mathbf{x}_N'\}$. Unlike in the case of pure camera rotation, this mapping not only depends upon the 3-D translation vector but also upon the actual 3-D location of each individual point observed. Therefore, τ is, in general, a non-linear mapping of the image (i.e., a *warping*) with $N+2$ degrees of freedom[1], whereas the rotational mapping in (3.9) has only 2 degrees of freedom (3 in the case of unrestricted motion).

Despite the complex image transformation caused by pure translation, one important property of τ can be described exclusively in the image plane: each displacement vector must lie a straight line passing through the original point and one unique location in the image (the FOE). This means that, if the vehicle is undergoing pure translation, then there must exist an image location $x_f = (x_f y_f)^T$ such that

$$\exists x_f \left\{ \forall \ (x_i, x_i') \in \tau : \ x_i' = x_i + \mu_i(x_i - x_f), \ \mu_i \in \mathbb{R}^+ \right\} . \tag{3.16}$$

If the condition in (3.16) holds, we say that τ is a *radial mapping* between the images I and I' with respect to the FOE x_f. The important fact here is that the displacement field created by pure camera translation (in a stationary environment) is always *radial*, no matter at what distance the observed points are located in 3-D.

3.5 COMPUTING THE TRANSLATION PARAMETERS

Figure 3.5 shows the geometric relationships for measuring the amount of camera translation in the 2-D case. It can be considered as a top view of the camera, i.e., a projection onto the X/Z-plane of the camera-centered coordinate system. The cross section of the image plane is shown as a straight line. The camera is translating from left to right in the direction given by $\mathbf{T} = (x_f f)^T$.

A stationary 3-D point is observed at two instances of time, which moves relative to the camera from $\mathbf{X}$ to $\mathbf{X}'$, where

---

[1] The 3-D distances Z_i of the N points being observed, plus the two coordinates of the FOE.

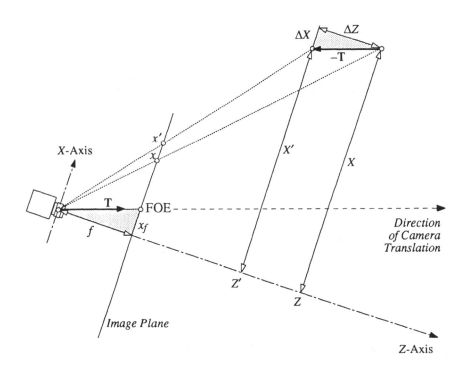

Figure 3.5 Expansion from the FOE. The camera moves by a vector **T** in 3-D space, which passes through the lens center and the FOE in the camera plane. The 3-D Z-axis coincides with the optical axis of the camera.

$$\mathbf{X} = \begin{bmatrix} X \\ Z \end{bmatrix} \quad \text{and} \quad \mathbf{X'} = \begin{bmatrix} X' \\ Z' \end{bmatrix} = \begin{bmatrix} X - \Delta X \\ Z - \Delta Z \end{bmatrix}, \qquad (3.17)$$

resulting in two image points x and x'. Using the inverse perspective transformation (3.8) yields

$$Z = \frac{f}{x}X \quad \text{and} \quad Z' = Z - \Delta Z = \frac{f}{x}X' = \frac{f}{x}(X - \Delta X). \qquad (3.18)$$

From similar triangles (shaded in Figure 3.5), we get

$$\frac{\Delta X}{x_f} = \frac{\Delta Z}{f} \, ,$$ (3.19)

and finally

$$Z = \Delta Z \frac{x' - x_f}{x' - x} = \Delta Z \left(1 + \frac{x - x_f}{x' - x} \right) \, .$$ (3.20)

Thus the rate of expansion of image points from the FOE contains direct information about the depth of the corresponding 3-D points. Consequently, if the vehicle is moving along a straight line and the FOE has been located, the 3-D structure of the scene can be determined from the expansion pattern in the image, at least up to an unknown scale factor. If, in addition, the velocity of the vehicle ($\Delta Z/\Delta t$) in space is known, the *absolute* distance of any stationary point can be determined. Alternatively, the linear velocity of the vehicle can be obtained if the true distance of one point in the scene is known (e.g., from laser range data) or if at least one coordinate value of a 3-D point is known (see Appendix, Section A.2). In practice, of course, any such 3-D depth computation requires that *(a)* the FOE can be located with sufficient accuracy and that *(b)* the observed image points exhibit a significant rate of divergence away from the FOE. As will be shown in Chapter 4, imaging noise and point location errors pose serious problems against meeting both criteria.

Having discussed the individual effects of camera rotation and translation in this chapter, we now turn to the problem of *decomposing* the image effects of arbitrary camera motion into their rotational and translational components. Our approach for doing this is based on two important relationships that have been worked out in this chapter. First, we have shown that the effects of pure camera rotations are independent of the 3-D scene structure and can therefore be removed if the rotation angles are known (3.10). The second observation was that pure camera translation always results in a radial displacement field (3.16), no matter where the observed points are actually located in 3-D space.

DECOMPOSING IMAGE MOTION

Under arbitrary camera motion, the resulting motion sequence is a superposition of the individual image effects of camera rotation and translation which we discussed in the previous chapter. Extracting the rotational and translational components from the observed displacement field is important for two reasons. First, they can be used to compute the camera's motion parameters which are useful for controlling the vehicle. The second motivation is that we can use the *translational* components of the displacement vector field to obtain information about the static 3-D scene structure and to determine the movements of non-stationary objects.

In the following, we develop methods for motion decomposition that are based on the concept of the *Focus of Expansion* (FOE). Section 4.1 presents the main elements of the FOE concept. Unfortunately, computing the FOE under arbitrary camera motion is a difficult problem. In Sections 4.2 and 4.3 we describe two approaches towards this goal that differ mainly by their underlying search strategy. In the *FOE-from-Rotations* approach (Section 4.2), we assume that the approximate camera rotations are given and then search for the location of the FOE. This assumption is not unreasonable, because camera rotations could be known from other sources. The alternative approach, termed *Rotations-from-FOE* (Section 4.3), assumes that the location of the FOE is known and the problem is to determine the corresponding camera rotations.

4.1 MOTION BETWEEN SUCCESSIVE FRAMES

As we have shown in the previous chapter, the image effects of camera rotations and translations in a stationary environment can be described as 2-D mappings $\rho_{\phi\theta}$ (3.10) and τ (3.16) respectively. Similarly, for the combined 3-D camera motion consisting of the rotations $\mathbf{R}_\phi$ and $\mathbf{R}_\theta$ and the translation $\mathbf{T}$, where a 3-D point $\mathbf{X} = (X\ Y\ Z)^T$ is transferred to $\mathbf{X}' = (X'\ Y'\ Z')^T$,

$$\mathbf{X} \to \mathbf{X}' = \mathbf{R}_\phi\ \mathbf{R}_\theta\ (\mathbf{T} + \mathbf{X}) , \qquad (4.1)$$

we define the image-space transformation $\delta(\mathbf{x})$ which expresses the image effects of this motion. δ is simply the concatenation ($\circ$) of the τ and $\rho_{\phi\theta}$ operators:

$$\mathbf{x} \to \mathbf{x}' = \rho_{\phi\theta}\ (\tau\ (\mathbf{x})) = \rho_{\phi\theta} \circ \tau\ (\mathbf{x}) = \delta(\mathbf{x}) . \qquad (4.2)$$

Figure 4.1 illustrates the consequences of combined translation and rotation. At two instances in time, t_0 and t_1, the position of the vehicle in space is specified by the location of a reference point $\mathbf{P}$ (i.e., the lens center of the camera) and the orientation Ω of the vehicle. The original image $\mathbf{I}_0$ is seen at time t_0.

The Focus of Expansion (FOE)

Following the adopted scheme of motion decomposition (Equation 4.1), the translation $\mathbf{T}$ is applied *first*, which takes the vehicle's reference point from position $\mathbf{P}_0$ to position $\mathbf{P}_1$, without changing its initial orientation Ω_0. At this point (position $\mathbf{P}_0$ and orientation Ω_0), the camera would see the intermediate image $\mathbf{I}^*$, which is the initial image $\mathbf{I}_0$ modified by the effects τ of the camera translation, i.e.

$$\mathbf{I}_0 \to \mathbf{I}^* = \tau\ (\mathbf{I}_0) . \qquad (4.3)$$

The FOE is located at the intersection of the 3-D translation vector $\mathbf{T}$ with the image plane $\mathbf{I}_0$. In the *second* step, the vehicle is rotated by ω to align with its final orientation Ω_1. At this point, image $\mathbf{I}_1$ is observed.

Notice that, unlike the two given images $\mathbf{I}_0$ and $\mathbf{I}_1$, the image $\mathbf{I}^*$ is never really observed, except in the special case of pure camera translation, i.e., when $\mathbf{I}^* = \mathbf{I}_1$. While it is not our goal to actually *compute* the image $\mathbf{I}^*$, we use it in the following to derive our decomposition algorithms. The unknown image $\mathbf{I}^*$ is related to the observed images $\mathbf{I}_0$ and $\mathbf{I}_1$ in two important ways. *First*, as we would expect from pure camera translation, the images $\mathbf{I}_0$ and $\mathbf{I}^*$ form a radial mapping (Equation 3.16) with respect to the (unknown) FOE $\mathbf{x}_f = (x_f\ y_f)^T$.

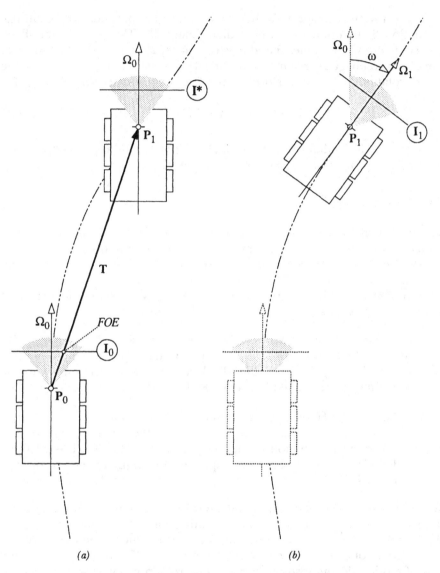

(a)   (b)

Figure 4.1 Arbitrary camera motion between two frames. Vehicle motion between the initial position (where image I_0 is observed) and the final position (where image I_1 is observed) is modeled as two separate steps. *(a)* First, the vehicle translates by a 3-D vector **T** from position P_0 to position P_1 without changing its orientation Ω_0. After this step, the intermediate image I^* would be seen. Subsequently *(b)*, the vehicle rotates by changing its orientation from Ω_0 to Ω_1. Now image I_1 is observed. The *Focus of Expansion* (FOE) is found where the vector **T** intersects the image plane I_0 (and also I^*).

Secondly, the final image I_1 is the result of a rotational transformation $\rho_{\phi\theta}$ (Equation 3.10) upon the intermediate image I^*. This means that, if the correct pan and tilt angles θ and ϕ were known, I^* could be obtained from the observed image I_1 (regardless of the 3-D scene structure) by applying the *inverse* rotational image transformation $\rho_{\phi\theta}^{-1}$ to the second observed image I_1:

$$I^* = \tau\,(I_0) = \rho_{\phi\theta}^{-1}\,(I_1)\,. \tag{4.4}$$

We call the elimination of the rotational components by applying $\rho_{\phi\theta}^{-1}$ to the image I_1, the *derotation* of the displacement field.

Two Search Strategies

The task of computing the camera motion parameters can thus be stated as a search problem in the four-dimensional space spanned by θ, ϕ, and $x_f = (x_f\,y_f)^T$. Equation 4.4 provides us with two powerful constraints upon the intermediate image I^* which we can use to compute the unknown motion parameters. In particular, the relation $\tau\,(I_0) = \rho_{\phi\theta}^{-1}(I_1)$ suggests two principal search strategies for separating the motion components:

(1) *FOE from Rotations:* Successively apply rotational mappings $\rho_{\phi_j\theta_j}^{-1}$, with $(\theta_j\,\phi_j) := (\theta_1\,\phi_1)$, $(\theta_2\,\phi_2)$,... to the second image I_1, until the resulting image $I_j^* = \rho_{\phi_j\theta_j}^{-1}\,(I_1)$ is a *radial* displacement field with respect to the original image I_0. Then locate the FOE x_f in I_i^*.

(2) *Rotations from FOE:* Successively select FOE-locations (different directions of vehicle translation) $x_{f_j} := x_{f_1}, x_{f_2},...$ in the original image I_0 and see if an inverse rotational mapping $\rho_{\phi_j\theta_j}^{-1}$ exists that makes $I_i^* = \rho_{\phi_j\theta_j}^{-1}(I_1)$ a radial displacement field with respect to the original image I_0 and the selected FOE x_{f_j}.

We have developed and evaluated algorithms for both decomposition approaches which we discuss in the remaining parts of this chapter. In reality, the relation in (4.4) usually cannot be satisfied due to the effects of discretization, noise, and other point location errors. This fact turns the problem of motion decomposition into an optimization problem with respect to some (yet to be defined) distance function. As it turns out, the crucial problem in the *FOE-from-Rotations* approach (Section 4.2) is to determine, if a displacement field is (or is close to being) radial when the location of the FOE is unknown. Measuring the deviation from a radial pattern is, in fact, much simpler when the location of the FOE is known, as in the *Rotations-*

from-FOE approach (Section 4.3). Consequently, the latter method has been extended to the "Fuzzy FOE" technique (see Chapter 5) which, in turn, has become the basis of our overall motion analysis approach.

Although there have been a number of suggestions for FOE-algorithms in the past [35, 41, 56, 59, 61], hardly any results on real outdoor imagery have been published. One reason for the absence of satisfactory results may be that most researchers have tried to locate the FOE in terms of a single, distinct image location. In practice, however, the noise generated by only digitizing a perfect translatory displacement field may keep the resulting vectors from passing through a single pixel. Generally, it appears that any method which relies on *extending* displacement vectors to find the FOE is inherently sensitive to image degradations. Interestingly, even humans seem to have difficulties in determining the exact direction of heading (i.e., the location of the FOE on the retina) under certain conditions. Average deviation of human judgement from the real direction has been reported from 1° to as large as 10° and up to 20° in the presence of large rotations [58, 74, 75, 76].

4.2 FOE FROM ROTATIONS

Given a displacement field defined by the two observed images I_0 and I_1, the main steps of this approach are:

(1) Guess the camera rotation angles θ_j and ϕ_j (for the current iteration *j*).

(2) Derotate the displacement field by θ_j, ϕ_j to obtain $I_j^* = \rho_{\phi_j\theta_j}^{-1}(I_1)$.

(3) Repeat steps *(1)* and *(2)* until I_k^* is a radial displacement field with respect to I_0.

(4) Locate the FOE in the final derotated image I_k^*.

First, the rotational components are estimated and their inverses are applied to "derotate" the image in the second step. If the rotation estimate were accurate, the derotated displacement field would diverge from a single image location (the FOE). The *third* step checks if the displacement field is radial, i.e., how much it deviates from a radial field. Finally, when the displacement field is acceptably radial, the location of the FOE is determined.

Thus three problems have to be solved in the *FOE-from-Rotations* approach:

(*a*) How to initially estimate the camera rotations?

(*b*) How to determine the "radialness" of a given displacement field?

(*c*) How to locate the FOE in an almost radial displacement field?

Initial Rotation Estimates

There are several alternatives to obtain the initial estimates for the camera rotations. Each 2-D vector in the displacement field is the sum of vector components caused by camera rotation and camera translation. Since the displacement caused by translation depends on the depth of the corresponding points in 3-D space (Equation 3.20), points located at a large distance from the camera are not significantly affected by camera translation. Therefore, one way of estimating vehicle rotation is to compute θ and ϕ from displacement vectors which are *known* to belong to points at far distance. Under the assumption that those displacement vectors are only caused by rotation, (3.14) can be applied to find the two angles. In some situations, distant points are selected easily. For example, points on the horizon are often located at a sufficient distance from the vehicle. Image points close to the axis of translation would be preferred because they diverge from the FOE slower than other points at the same depth.

Of course, points at far distances may not always be available or may not be *known* to exist in the image. In the Appendix of this book (Section A.1), we describe a technique for estimating the camera motion parameters (translation *and* rotations) by using incremental geometric constraints, without the use of distant 3-D points. In some applications, information about camera rotations may also be available from other sensors, such as inertial navigation sensors (gyros and accelerometers) that are attached to the camera platform [62].

The initial estimate for the camera rotations is not really critical. If the time between successive frames is reasonably small, the computed rotations for one frame pair can be used as the initial value for the subsequent frame. Knowledge of the camera's angular acceleration can make this estimate even more accurate.

Measuring "Radialness" and Locating the FOE

After applying a certain derotation to the displacement field, the question is how close the new displacement field is to a *radial mapping*, where all vectors diverge from one image location. If the displacement field is perfectly radial, then only the displacement components due to camera translation remain. In the following, we discuss two different methods for measuring "radialness" and for locating the FOE. The *first* method intersects the displacement vectors with horizontal and vertical straight lines and uses the distribution of the intersections on these lines to measure "radialness". The second method is similar in principle, but uses the intersections of displacement vectors with *two* parallel lines in both horizontal and vertical directions to measure the "radialness" of the displacement field.

Method 1: Intersections with a Single Straight Line

Prazdny [56] suggested to estimate the disturbance of the displacement field by computing the variance of intersections of one displacement vector with all other vectors. If the intersections lie in a small neighborhood, then the variance is small, which indicates that the displacement field is almost radial.

The problem can be simplified by intersecting the displacement vectors with imaginary horizontal and vertical lines, whose orientation is not affected by different camera rotations [21]. Figure 4.2 shows five displacement vectors $P_1 \to P_1'$,... $P_5 \to P_5'$ that intersect a vertical line positioned at x_v through points y_1 ...y_5. Moving the vertical line from x_v towards x_0 will bring the points of intersection closer together and will thus result in a smaller variance. The point of intersection of a displacement vector $P_i \to P_i'$ with a vertical line at position x_v is given by $(x_v\, y_i)^T$, where

$$y_i = \frac{x_v(y_i - y_i') + x_i\, y_i' - x_i' y_i}{x_i - x_i'} \; . \tag{4.5}$$

The variance of intersection of all displacement vectors with the vertical line at position x_v is

$$\sigma_y^2(x_v) = \frac{1}{N}\left[\sum_{i,\, x_i \neq x_i'} y_i^2 - \frac{1}{N}\left(\sum_{i,\, x_i \neq x_i'} y_i \right)^2 \right], \tag{4.6}$$

where N is the number of displacement vectors included in the summation.

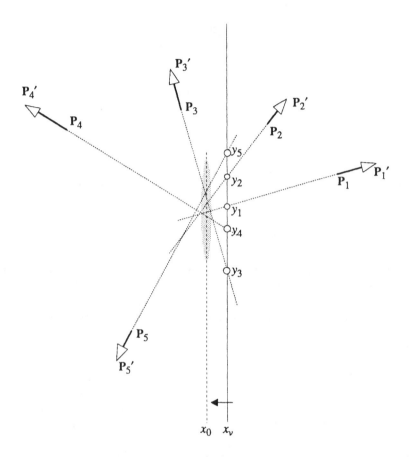

Figure 4.2 Intersecting the displacement vectors with a vertical line. When the vertical line at horizontal position x_v is moved towards x_0, the points of intersection move closer together, and therefore, the *variance* of the y-coordinates y_i of the intersection points decreases.

To find the vertical cross section with minimum intersection variance, the first derivative of 4.6 with respect to x_v is set to zero. The *location* x_0 of minimum intersection variance is then obtained as

$$x_0 = 2 \frac{\frac{1}{N} \sum \left(\frac{y_i - y_i'}{x_i - x_i'}\right) \cdot \sum \left(\frac{x_i y_i' - x_i' y_i}{x_i - x_i'}\right) - \sum \left(\frac{(y_i - y_i')(x_i y_i' - x_i' y_i)}{(x_i - x_i')^2}\right)}{\sum \left(\frac{y_i - y_i'}{x_i - x_i'}\right)^2 - \frac{1}{N}\left(\sum \frac{y_i - y_i'}{x_i - x_i'}\right)^2}. \qquad (4.7a)$$

Similarly, the position y_0 of the horizontal cross section with minimal intersection variance is

$$y_0 = 2 \frac{\frac{1}{N} \sum \left(\frac{x_i - x_i'}{y_i - y_i'}\right) \cdot \sum \left(\frac{x_i' y_i - x_i y_i'}{y_i - y_i'}\right) - \sum \left(\frac{(x_i - x_i')(x_i' y_i - x_i y_i')}{(x_i - x_i')^2}\right)}{\sum \left(\frac{x_i - x_i'}{y_i - y_i'}\right)^2 - \frac{1}{N}\left(\sum \frac{x_i - x_i'}{y_i - y_i'}\right)^2}. \qquad (4.7b)$$

The range of summation in (4.7) is the same as in (4.6).

Locating the FOE

For a given displacement field, we can use (4.7) to compute the vertical and horizontal cross sections through the extended displacement vectors that have a minimum "width". Under ideal conditions, the true FOE would be located exactly at position $(x_0 \ y_0)$. Of course, normally we cannot expect the derotated displacement field to be perfectly radial, either due to imperfect rotation estimates or due to noise and location errors.

In the following, we evaluate the formulation in (4.7) under various conditions, using the synthetic displacement field shown in Figure 4.3 which is also used for the subsequent experiments. The square around the center (± 100 pixels in both directions) marks the region over which the evaluation was performed. In Figures 4.4–4.7, the *standard deviation* $s_y(x_v)$, i.e., the square root of the intersection variance in (4.6) has been evaluated under *(1)* varying residual rotation in vertical direction and *(2)* varying levels of noise.

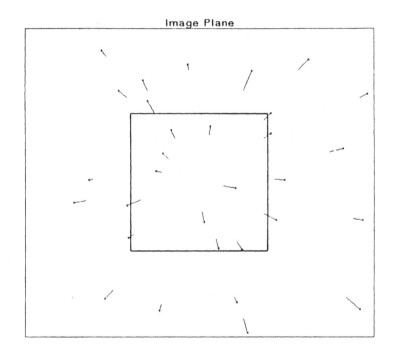

Figure 4.3 Displacement field used to evaluate various error functions. The square in the center (±100 pixels in both directions) outlines the region of evaluation.

Figure 4.4 shows the distribution of $s_y(x_v)$, under noiseless conditions Locations of displacement vectors are represented by real numbers, i.e., not rounded to integer values. In Figure 4.4a, no residual rotation exists ($\theta = \phi = 0$), i.e., the displacement field is perfectly radial. The value of the horizontal position x_v of the cross section varies ±100 pixels around the true FOE. $s_y(x_v)$ is zero for $x_v = x_f$ (the x-coordinate of the FOE) and increases linearly on both sides of the FOE. In Figures 4.4b–d, the vertical rotation ϕ is increased from 0.2° to 1.0°, the horizontal rotation θ is zero. The bold vertical bars mark the horizontal position of minimum standard deviation, the thin bars mark the location of the FOE. It can be seen in Figure 4.4 that the *amount* of minimum standard deviation rises with increasing vertical rotation, but that the *location* of minimum variance remains in close proximity of the true FOE.

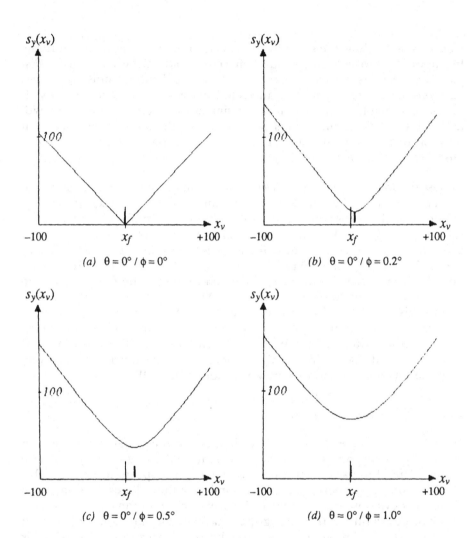

Figure 4.4 Standard deviation of straight-line intersections – noiseless case. Displacement vectors are intersected by a vertical straight line whose horizontal position x_v varies in the range of ±100 pixels (abscissa) around the actual FOE. The ordinate in each plot *(a–d)* shows the *standard deviation* $s_y(x_v)$, of the intersection points' y-coordinates. The vertical camera rotation ϕ is varied from $0°$ *(a)*, $0.2°$ *(b)*, $0.5°$ *(c)*, to $1.0°$.*(d)* The horizontal camera rotation θ is $0°$ in *(a–d)*. No noise was applied and image positions were assumed to be real numbers. The horizontal position of the true FOE ($x_f = 0$) is indicated by thin marks. Bold marks have been placed at the locations of *minimum* standard deviation.

Figures 4.5–4.7 show the same function $s_y(x_v)$, under the influence of noise. In Figure 4.5, noise was applied by merely rounding the locations of displacement vectors to their nearest integer values. Uniform noise of ±1 and ±2 pixels was added to image locations in Figures 4.6 and 4.7 respectively. It can be seen that the effects of noise are similar to the effects caused by residual vertical rotation components. It thus appears that finding the horizontal position of the FOE through the intersection variance is possible under moderate *vertical* rotations and noise.

To test the influence of residual *horizontal* rotation, we have plotted in Figure 4.8 the value x_0, i.e., the *location* of minimum $s_y(x_v)$, for all x_v, under varying horizontal rotation θ. The vertical rotation ϕ is kept fixed in each of the four plots in Figures 4.8a–d. The horizontal camera rotation θ (−1° to +1°) is shown on the abscissa. The ordinate gives the location of minimum variance x_0 in the range of ±100 pixels around the FOE (marked x_f). If the location x_0 were invariant to residual horizontal rotations, the functions in Figure 4.8 would all be horizontal lines at x_f. Obviously, there is no such invariance. This is not surprising since the effects of horizontal camera rotation are similar to horizontal translation parallel to the image plane. The conclusion is that, with this formulation, we can locate the FOE only in a displacement field that is already highly radial. In other words, we need very accurate rotation estimates before we can locate the FOE.

Improving the Rotation Estimates

In order to improve the rotation estimates, we must search for rotation angles that better derotate the displacement field, i.e., they make it more radial. The minimal intersection variance $\sigma_y^2(x_0)$ appears to be a suitable measure for this property. Figures 4.9 and 4.10 show the evaluation of the standard deviation $s_y(x_0)$ under residual vertical rotations and noise. In order to be useful for recovering the rotations, the function plots should have at least a local minimum at the true horizontal rotation (zero), which is marked by a vertical line in the center. This is obviously the case in Figures 4.9a and 4.10a. However, even in these cases (where the vertical rotation is zero) there are local minima in the neighborhood of the true value where a local search algorithm would get trapped. This becomes worse with increasing vertical rotation, as shown in Figures 4.9b–d and 4.10b–d, although the amount of rotation is only moderate (±1.0°).

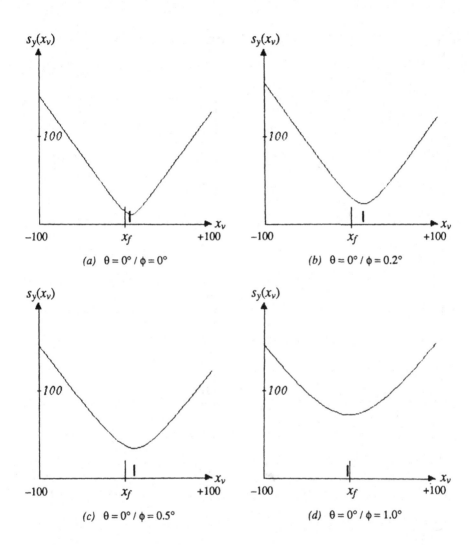

Figure 4.5 Standard deviation of straight-line intersections – effects of spatial quantization. The setup is identical to the one in Figure 4.4, except that displacement vector coordinates were rounded to closest integer values. No additional noise was applied. The vertical camera rotations ϕ are again 0° *(a)*, 0.2° *(b)*, 0.5° *(c)*, and 1.0° *(d)*. The horizontal position of the true FOE ($x_f = 0$) is indicated by thin marks.

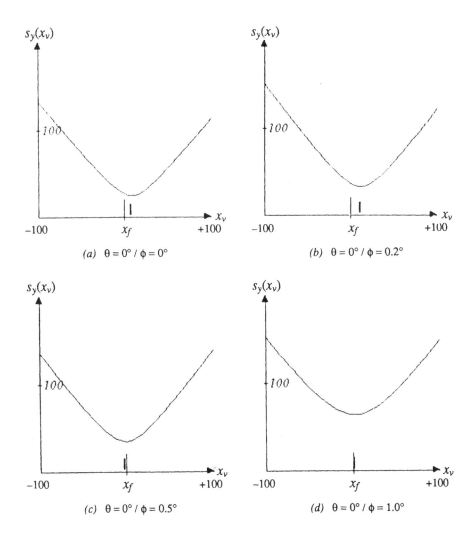

Figure 4.6 Standard deviation of straight-line intersections – effects of random noise (±1 pixel). The setup is identical to the one in Figure 4.4, except that uniformly distributed random noise in the range of ±1 pixel was applied to the image locations before quantization. The vertical camera rotations ϕ are again $0°$ *(a)*, $0.2°$ *(b)*, $0.5°$ *(c)*, and $1.0°$ *(d)*. The horizontal position of the true FOE $(x_f = 0)$ is indicated by thin marks.

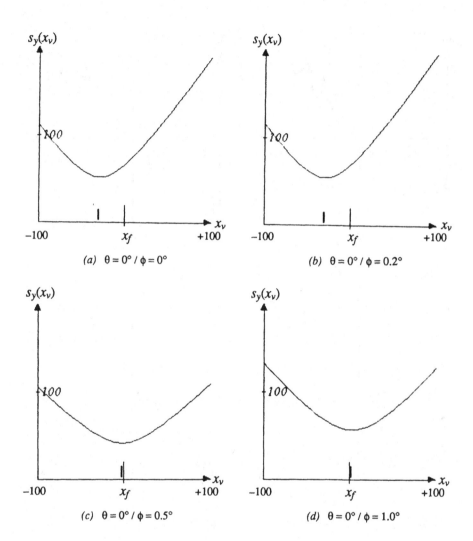

Figure 4.7 Standard deviation of straight-line intersections – effects of random noise (±2 pixels). The setup is identical to the one in Figure 4.4, except that uniformly distributed random noise in the range of ±2 pixels was applied to the image locations before quantization. The vertical camera rotations ϕ are again 0° (a), 0.2° (b), 0.5° (c), and 1.0° (d). The horizontal position of the true FOE ($x_f = 0$) is indicated by thin marks.

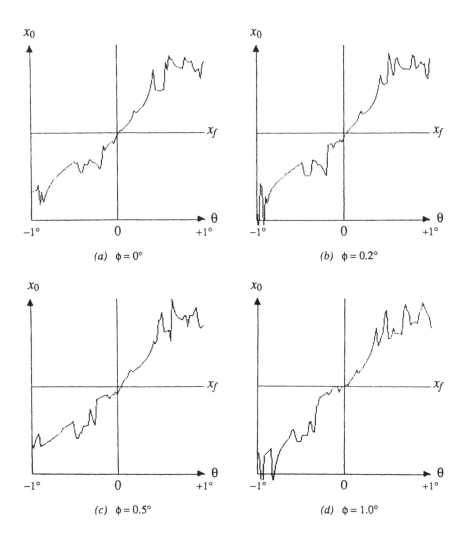

Figure 4.8 *Location* of minimum intersection standard deviation. The ordinate in each plot *(a–d)* shows the location of minimum intersection standard deviation (relative to the actual FOE located at x_f) under varying horizontal rotations θ in the range of $\pm1°$ (the vertical line in the center marks zero rotation). The vertical camera rotation ϕ is kept fixed for each plot, taking the values $0°$ *(a)*, $0.2°$ *(b)*, $0.5°$ *(c)*, to $1.0°$.*(d)*. Image locations were digitized but no additional noise was added. Ideally, the location of minimum intersection standard deviation should always be close to the true FOE, i.e., each plot should consist of a horizontal line at x_f.

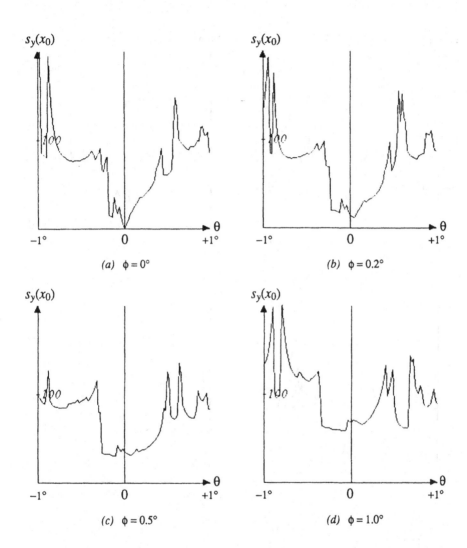

Figure 4.9 *Values* of minimum intersection standard deviation – noiseless case. The ordinate in each plot *(a–d)* shows the value of minimum intersection standard deviation under varying horizontal rotations θ in the range of $\pm 1°$ (the vertical line in the center marks zero rotation). The vertical camera rotation ϕ is kept fixed for each plot, taking the values $0°$ *(a)*, $0.2°$ *(b)*, $0.5°$ *(c)*, to $1.0°$ *(d)*. Image locations were discretized but no additional noise was added.

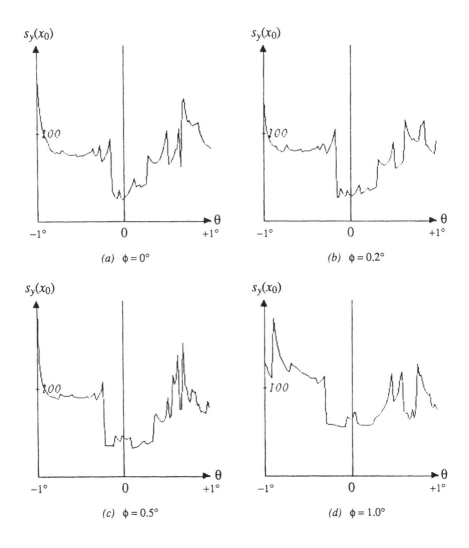

Figure 4.10 *Values* of minimum intersection standard deviation – effects of random noise (±2 pixels). The setup is identical to the one in Figure 4.9, except that uniformly distributed random noise in the range of ±2 pixels was applied to the image locations before quantization. The vertical camera rotations φ are again 0° *(a)*, 0.2° *(b)*, 0.5° *(c)*, and 1.0°*(d)*.

In summary, we can use the intersections of displacement vectors with a single line to locate the FOE in a displacement field that is almost radial. We would then use two orthogonal lines to determine both coordinates of the FOE. However, due to the resulting functions being locally non-convex, the formulation appears to be unsuitable for evaluating a displacement field with respect to its "radialness" under realistic conditions, which would be necessary to find good rotation estimates.

Method 2: Intersection with Two Parallel Lines

The second method in the *FOE-from-Rotations* category again uses the points of intersection at vertical (and horizontal) lines. The basic idea is illustrated in Figure 4.11. In contrast to the previous scheme (Method 1), the displacement vectors are intersected by *two* vertical lines, both of which lie on the same side of the FOE. Since the location of the FOE is not known, the two lines are simply located at a sufficient distance away from any possible FOE-location. This results in two sets of intersection points $\{(x_1\ y_{1i})\}$ and $\{(x_2\ y_{2i})\}$. If all displacement vectors emanate from one single image location, then the distances between corresponding intersection points in the two sets must be proportional, i.e.,

$$\frac{y_{1i}-y_{1j}}{y_{2i}-y_{2j}} = \frac{y_{1j}-y_{1k}}{y_{2j}-y_{2k}} \tag{4.8}$$

for all i, j, k. Therefore, a linear relationship exists between the vertical co-ordinates of intersection points on these two lines. The "goodness" of this linear relationship is easily measured by computing the linear correlation coefficient $r_{1,2}$ for the y-coordinates of the two sets of points:

$$r_{1,2} = \frac{\sum y_{1i}\,y_{2i} - \frac{1}{N}\left(\sum y_{1i} \cdot \sum y_{2i}\right)}{\left[\sum y_{1i} - \frac{1}{N}\left(\sum y_{1i}\right)^2\right]^{1/2}\left[\sum y_{2i} - \frac{1}{N}\left(\sum y_{2i}\right)^2\right]^{1/2}} \tag{4.9}$$

$r_{1,2}$ is a real number in the range $[-1, +1]$. If both vertical lines are on the same side of the FOE, then the optimal value is $+1$. Otherwise, if the FOE lies between the two lines, the optimal coefficient is -1. The horizontal position of the two vertical lines is of no importance, as long as one of these conditions is satisfied. For example, the left and right border lines of the image can be used.

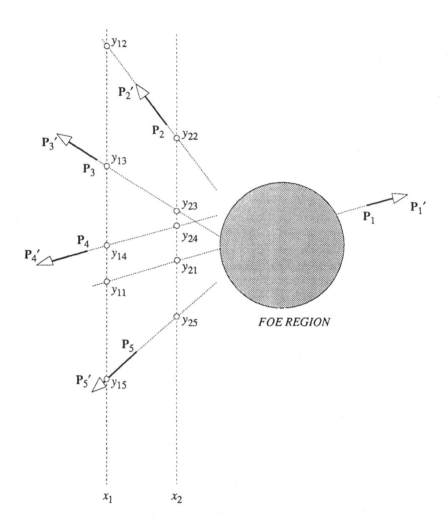

Figure 4.11 Intersecting displacement vectors with a parallel line pair located on one side of the FOE.

Figures 4.12 and 4.13 show plots for this correlation coefficient under the same conditions as in Figures 4.9 and 4.10. No noise was applied for Figure 4.12. The shapes of the curves are similar to those for the minimum standard deviation shown earlier, with peaks at the same locations. It is apparent, however, that each curve has several locations where the coefficient is close to the optimum value (+1), i.e., no distinct global optimum exists. The same is true when noise is present (Figure 4.13). This behavior makes the method of maximizing the correlation coefficient practically useless for computing the FOE.

The two methods we have presented in connection with the *FOE-from-Rotations* approach are similar in that they both use a linear statistical measure for evaluating the radialness of the displacement field and for locating the FOE. The *variance* of point coordinates was used in *Method 1* and the *linear correlation coefficient* in *Method 2*. Of course, one could employ more sophisticated techniques for solving this optimization problem. It seems, however, that the basic problem of this approach is not the solution method but rather the way of measuring the deviation from the optimal (i.e., radial) displacement field. In particular, the idea of *extending* displacement vectors (which may be quite short) and using their intersections does not appear to be a good concept because any noise and discretization errors are effectively multiplied. As a consequence, we did not incorporate the *FOE-from-Rotations* approach in the final DRIVE system, but instead used a version of the *Rotations-from-FOE* approach, which is described in the following section.

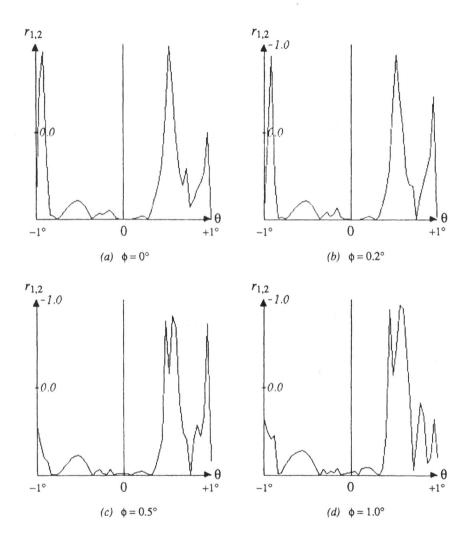

Figure 4.12 Correlation of intersections with parallels – noiseless case. Displacement vectors are intersected by a pair of parallel vertical lines. The ordinate in each plot *(a–d)* shows the *linear correlation coefficient* for the intersection points' *y*-coordinates on each line under varying horizontal rotations θ in the range of $\pm 1°$ (abscissa). The vertical camera rotation ϕ is kept fixed for each plot, taking the values $0°$ *(a)*, $0.2°$ *(b)*, $0.5°$ *(c)*, to $1.0°$ *(d)*. A perfectly radial displacement field has a coefficient of $+1.0$ (horizontal axis).

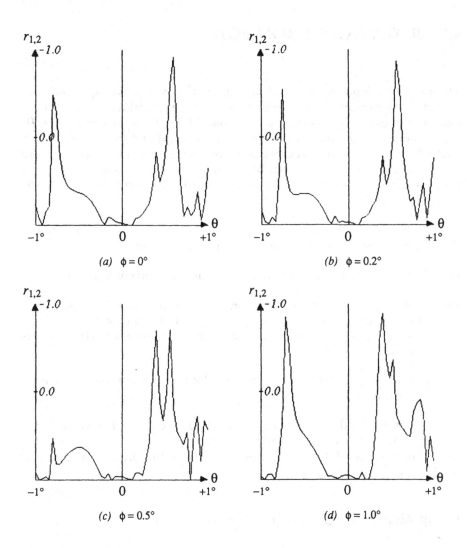

Figure 4.13 Correlation of intersections with parallels – effects of random noise (±2 pixels). The setup is identical to the one in Figure 4.12, except that uniformly distributed random noise in the range of ±2 pixels was applied to the image locations before quantization. The vertical camera rotations ϕ are again 0° *(a)*, 0.2° *(b)*, 0.5° *(c)*, and 1.0°*(d)*.

4.3 ROTATIONS FROM FOE

The main problem encountered with the *FOE-from-Rotations* approach presented in the previous section was to measure the "radialness" of a given displacement field with respect to an unknown FOE. Disturbances induced by
noise and residual rotation components are amplified by extending relatively short displacement vectors and computing their intersections. The
Rotations-from-FOE method described in this section avoids this problem by
picking an FOE-location first and then estimating the optimal derotation for
that particular FOE .

Given a displacement field defined by the two point sets corresponding to
the images I_0 and I_1, the main steps of this approach are:

(1)   Guess an FOE-location $x_f^{(j)}$ in image I_0 (for the current iteration j).

(2)   Determine the rotation mapping $\rho_{\phi,\theta_j}^{-1}$ such that $I_j^* = \rho_{\phi,\theta_j}^{-1}(I_1)$ is "most
        radial" with respect to the original image I_0 and the selected FOE $x_f^{(i)}$
        (3.16). Let $E^{(j)}$ measure the deviation of I_j^* from a perfectly radial
        pattern.

(3)   Repeat steps *(1)* and *(2)* until an FOE-location $x_f^{(k)}$ with minimum error
        $E^{(k)}$ is found.

An initial guess for the FOE-location can be obtained from knowledge about
the orientation of the camera with respect to the vehicle. For the subsequent frames, the FOE-location computed from the previous pair can be
used as a starting point.

Computing the Angles of Rotation

Once a particular x_f has been selected, the problem is to compute the rotation mappings $\rho_{\phi\theta}^{-1}$ which, when applied to the image I_1, will result in an
optimal radial mapping with respect to I_0 and x_f.

To measure how close a given mapping is to a radial mapping, the distances
between the points in the second image (x_i') and the "ideal" displacement
vectors are measured. The "ideal" displacement vectors lie on straight lines
passing through the FOE x_f and the points in the first image $x_i \in I_0$ (Figure

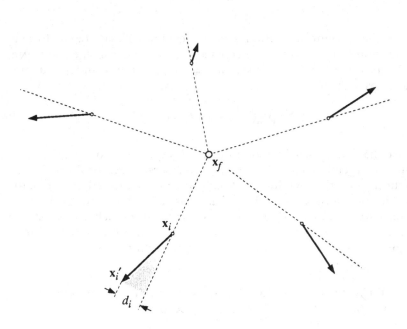

Figure 4.14 Measuring the deviation from a radial displacement field. For the assumed FOE-location $\mathbf{x}_f$, d_i is the perpendicular distance between the end point of the displacement vector $\mathbf{x}_i \rightarrow \mathbf{x}_i'$ and the radial line that originates from $\mathbf{x}_f$, and passes through $\mathbf{x}_i$. The sum of the squared distances d_i^2 is used to quantify the deviation from a radial displacement field with respect to $\mathbf{x}_f$.

4.14). The sum of the squared perpendicular distances d_i is used as the error measure. For a given set of corresponding image points ($\mathbf{x}_i \in I_0$, $\mathbf{x}_i' \in I_1$), the error measure is defined as

$$E(\mathbf{x}_f) = \sum_i E_i = \sum_i d_i^2 = \sum_i \left\{ \frac{1}{\|\mathbf{x}_i - \mathbf{x}_f\|^2} \left[(\mathbf{x}_i - \mathbf{x}_f) \times (\mathbf{x}_i' - \mathbf{x}_f) \right] \right\} \qquad (4.10)$$

Notice that this expression implicitly puts more weight upon the long (i.e., dominant) displacement vectors and less weight on the short ones.

Approximating Small Rotations

Since under perspective transformation, image points move along hyperbolic paths when the camera performs pure rotation, the resulting displacement is not uniform over the entire image plane (3.11, 3.12). However, if the amount of rotation is small (less than 4°, say) or the focal length of the camera is large, we can replace the nonlinear image transformation $\rho_{\phi\theta}^{-1}$ by a uniform, linear translation $s_{\phi\theta} = (s_x \ s_y)^T$ that is independent of the image position, i.e.,

$$\mathbf{I}^* = \rho_{\phi\theta}^{-1}(\mathbf{I}_1) \approx s_{\phi\theta} + \mathbf{I}_1 = \mathbf{I}^+ . \tag{4.11}$$

Notice that the approximation in (4.11) is equivalent to assuming an orthographic projection model instead of the perspective projection. In most practical cases, the condition of small rotations is satisfied, provided that the time interval between successive frames is sufficiently small. However, the simplification in (4.11) does not imply that the approach works just for small rotations. In the case of large rotations, we can incrementally *derotate* the image, thus successively improving the linear shift approximation until the rotations vanish.

Using the approximation in (4.11), the original error function (4.10) for two corresponding images $\mathbf{I}_0$ and $\mathbf{I}_1$ now becomes

$$E_s(\mathbf{x}_f) = \sum_i \left\{ \frac{1}{\|\mathbf{x}_i - \mathbf{x}_f\|^2} \left[(\mathbf{x}_i - \mathbf{x}_f) \times (\mathbf{x}_i' - \mathbf{x}_f + \mathbf{s}) \right] \right\} \tag{4.12}$$

$$= \sum_i \frac{\left[(x_i - x_f)(y_i' - y_f + s_y) - (y_i - y_f)(x_i' - x_f + s_x) \right]^2}{(x_i - x_f)^2 + (y_i - y_f)^2}$$

where $\mathbf{x}_i \in \mathbf{I}_0$, $\mathbf{x}_i' \in \mathbf{I}_1$, and $\mathbf{s} = s_{\phi\theta}$. For a given FOE-location $\mathbf{x}_f$, the problem is to minimize $E_s(\mathbf{x}_f)$ with respect to the two unknowns s_x and s_y. This is a least-squares problem that can be solved with standard numerical techniques.

Using a Pivot Point

To reduce this problem to a one-dimensional search, one privileged point $\mathbf{x}_g$, called the *pivot point*, is selected in image $\mathbf{I}_0$ and forced to maintain zero error for its displacement vector $\mathbf{x}_g \to \mathbf{x}_g'$ (Figure 4.15). In other words, we require that $\mathbf{x}_g^+ = \mathbf{s} + \mathbf{x}_g'$ (which is $\mathbf{x}_g'$ shifted by $\mathbf{s}$) must lie on a straight line passing through $\mathbf{x}_f$ and $\mathbf{x}_g$. Any shift $\mathbf{s}$ applied to the image $\mathbf{I}_1$ must keep $\mathbf{x}_g^+$ on this straight line, i.e.,

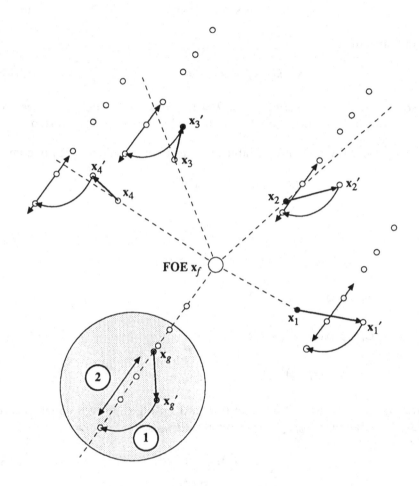

Figure 4.15 Use of a pivot point for finding the optimal image registration. The problem is to register the two images $I_0 = \{x_i\}$ and $I_1 = \{x_i'\}$ onto a radial pattern out of the FOE x_f (dashed lines) by shifting the second image I_1 uniformly by some (unknown) 2-D vector s. We can simplify this to a 1-D search by selecting one privileged displacement vector $x_g \to x_g'$ whose deviation from the radial line is forced to zero under any image translation s (x_g is called the *pivot point*). Intuitively, this means that the image I_1 is first shifted such that x_g' coincides with the radial line through x_g (1) and then all points in I_1 are translated parallel to that line (2) until a global minimum error $E_s(x_f)$ is reached.

$$\mathbf{x}_g^+ = \mathbf{s} + \mathbf{x}_g{}' = \mathbf{x}_f + \lambda(\mathbf{x}_g - \mathbf{x}_f) \tag{4.13}$$

for all **s**, and thus

$$\mathbf{s} = \mathbf{x}_f - \mathbf{x}_g{}' + \lambda(\mathbf{x}_g - \mathbf{x}_f) \tag{4.14}$$

for some $\lambda \in \mathbb{R}$. For example, for $\lambda = 1$, $\mathbf{s} = \mathbf{x}_g - \mathbf{x}_g{}'$ which is the vector $\mathbf{x}_g{}' \rightarrow \mathbf{x}_g$. In this case, the image $\mathbf{I}_1$ is shifted such that $\mathbf{x}_g$ and $\mathbf{x}_g^+$ overlap.

This leaves λ as the only free variable and the error function (4.12) becomes

$$E_\lambda(\mathbf{x}_f) = \sum_i \left[\lambda A_i + B_i - C_i \right]^2 \tag{4.15}$$

with

$$A_i = \frac{1}{r_i} \left[(x_g - x_f)(y_i - y_f) - (x_i - x_f)(y_g - y_f) \right]$$

$$B_i = \frac{1}{r_i} (x_i{}' - x_g{}')(y_i - y_f)$$

$$C_i = \frac{1}{r_i} (x_i - x_f)(y_i{}' - y_g{}')$$

$$r_i = \sqrt{(x_i - x_f)^2 + (y_i - y_f)^2} \; .$$

Differentiating (4.15) with respect to λ and setting the result equal to zero yields the parameter λ_{opt} for the optimal shift $\mathbf{s}_{opt}$ as

$$\lambda_{opt}(\mathbf{x}_f) = \frac{\sum A_i C_i - \sum A_i B_i}{\sum A_i^2} \; . \tag{4.16}$$

The optimal shift $\mathbf{s}_{opt}$ can now be computed using (4.14) and (4.16). However, to evaluate a selected FOE-location $\mathbf{x}_f$, we are primarily interested in the corresponding minimum error $E_{min}(\mathbf{x}_f)$ that would result from applying the approximate derotation $\mathbf{s}_{opt}$. It is obtained from (4.15) and (4.16) as

$$E_{min}(x_f) = \lambda_{opt}^2 \sum A_i^2 + 2\lambda_{opt}\left(\sum A_iB_i - \sum A_iC_i\right) - 2\sum B_iC_i + \sum B_i^2 + \sum C_i^2$$

$$= -\frac{\left(\sum A_iB_i - \sum A_iC_i\right)^2}{\sum A_i^2} - 2\sum B_iC_i + \sum B_i^2 + \sum C_i^2 \ . \tag{4.17}$$

The normalized error E_η shown in the results that follow (Figures 4.17–4.23) is defined as

$$E_\eta(x_f) = \sqrt{\frac{1}{N} E_{min}(x_f)} \ , \tag{4.18}$$

where N is the number of displacement vectors used for computing the FOE. Obviously, the error value $E_\eta(x_f)$ can be computed at least in time proportional to N.

Prohibited FOE-Locations

Since in a displacement field caused by pure forward camera translation all vectors must point *away* from the FOE, this restriction must hold for any candidate FOE-location. If, after applying $s_{opt}(x_f)$ to the second image I_1, the resulting displacement field contains vectors pointing *toward* the hypothesized x_f, then this FOE-location is *prohibited* and can be discarded from further consideration.

Figure 4.16 shows a field of five displacement vectors. The optimal shift s_{opt} for the given x_f is shown as a vector in the lower right-hand corner. When s_{opt} is applied to the end point x_1', the resulting modified displacement vector $x_1 \rightarrow x_1^+$ does not point away from the FOE, because its projection onto the line $\overline{x_f x_1}$ is a vector that points *toward* the FOE. This is not consistent with the radial expansion pattern expected under forward camera motion and, therefore, x_f is not a feasible FOE for the given displacement field. Of course, this only holds when all displacement vectors used for the FOE computation originate from stationary environmental points. A moving object may create arbitrary displacement vectors that may not be radial or point away from the FOE. In fact, this behavior is important for the detection of moving objects, as we shall see in Chapter 6.

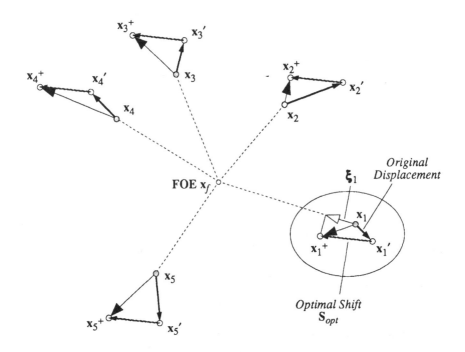

Figure 4.16 Prohibited FOE-locations. FOE-locations are *prohibited* if the displacement field resulting from the application of the optimal shift s_{opt} contains any vectors pointing toward the FOE. In the example shown above, this is true for point x_1. Shifting the end point x_1' by s_{opt} creates a modified displacement vector $x_1 \rightarrow x_1^+$. If this vector is projected onto the radial line $\overline{x_f\,x_1}$, the resulting vector ξ_1 points *toward* the FOE x_f. Assuming that all displacement vectors originate from stationary points, a single such event is sufficient to make the FOE-location x_f unfeasible (prohibited).

Evaluating a Single FOE

The following procedure **Evaluate–Single–FOE** examines one hypothetical FOE-location x_f for a given pair of images I_0 and I_1. It uses the function **Optimal–Shift** for computing the optimal shift vector s_{opt} and the corresponding error value $E_\eta(x_f)$. The function **Equivalent–Rotation** is used to obtain the rotation angles (3.14) equivalent to s_{opt}. The values of the constants ϕ_{max} and θ_{max} are the maximum angles of rotation for which the approximation in (4.11) is acceptable. If any of the estimated rotation angles ϕ^+ and θ^+ exceeds this limit, the intermediate image I^+ is derotated exactly (3.9) by procedure **Derotate–Image** before another estimate is computed. For sufficiently small rotations, the algorithm terminates after a single iteration. *Evaluate–Single–FOE* returns four values: the derotated image I^+, the estimated angles of camera rotation ϕ and θ, and the resulting normalized error $E_\eta(x_f)$:[1]

Evaluate–Single–FOE (x_f, I_0, I_1):
   $I^+ := I_1$;
   $(\phi, \theta) := (0, 0)$;
   repeat   */*usually only 1 iteration required*/*
      $(s_{opt}\ E_\eta) := $ *Optimal–Shift* (I_0, I^+, x_f);
      $(\phi^+, \theta^+) := $ *Equivalent–Rotation* (s_{opt});
      $(\phi, \theta) := (\phi, \theta) + (\phi^+, \theta^+)$;
      $I^+ := $ *Derotate–Image* (I_1, ϕ, θ);
   until $(|\phi^+| \leq \phi_{max}\ \&\ |\theta^+| \leq \theta_{max})$;
   return $(I^+, \phi, \theta, E_\eta)$.

Evaluating the Error Function

A first set of experiments was conducted on synthetic imagery to investigate the behavior of the error measure under various conditions, namely

- the average length of the displacement vectors (longer displacement vectors lead to a more accurate estimate of the FOE),

- the amount of residual rotation components in the image, and

- the amount of noise applied to the location of image points.

Figure 4.17 shows the distribution of the normalized error $E_\eta(x_f)$ for a sparse and relatively short displacement field containing seven vectors. Residual rotation components of ±2° in horizontal and vertical direction are present in Figure 4.17*b–d* to visualize their effects upon the image. This displacement field was used with different average vector lengths (indicated as *length-factor* = 2.0 = 8 pixels) for the other experiments on synthetic data. The displacement vector through the pivot point is marked with a heavy line.

---

[1] The pseudocode notation that we use here allows functions to return multiple values (as in Common LISP), and also allows multiple assignments, e.g.
    $(x,y,z) := $ *FOO*(a,b);
    $(u,v) := (r,s)$; .

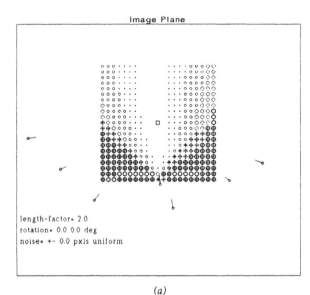

(a)

Image Plane

length-factor= 2.0
rotation= 0.0 0.0 deg
noise= +- 0.0 pxls uniform

Image Plane

length-factor= 2.0
rotation= 2.0 0.0 deg
noise= +- 0.0 pxls uniform

(b)

Figure 4.17 Displacement field and minimum error values for selected FOE-locations. The computed error values are plotted for FOE locations in an area of ±100 pixels around the true FOE (marked with a small square). The diameter of the circles drawn at each hypothesized FOE-location indicate the amount of normalized error (Equation 4.18), i.e., large circles are locations with large errors. Bold circles indicate error values above a certain threshold (set here to 5.0), *prohibited* locations (as defined earlier) are marked with ⊕. The amount of residual rotation varies: *(a)* no residual rotation; *(b)* 2.0° of horizontal rotation (camera rotated to the left).

	$E_\eta(\mathbf{x}_f) =$
□	0.0 – 1.0
·	1.0 – 2.0
∘	2.0 – 3.0
⊙	3.0 – 4.0
○	4.0 – 5.0
⬤	> 5.0
⊕	*prohibited*
▫	real FOE

Legend

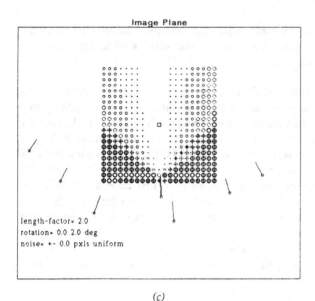

(c)

Figure 4.17 (*continued*) Residual rotations are: (*c*) 2.0° vertical rotation (camera rotated upward); (*d*) −2.0° vertical rotation (camera rotated downward).

Image Plane

length-factor= 2.0
rotation= 0.0 -2.0 deg
noise= +- 0.0 pxls uniform

(d)

	$E_\eta(x_f) =$
	0.0 – 1.0
·	1.0 – 2.0
∘	2.0 – 3.0
○	3.0 – 4.0
○	4.0 – 5.0
◉	> 5.0
⊕	*prohibited*
▫	real FOE

Legend

The choice of the pivot point is not critical but it should be located at a considerable distance from the FOE to reduce the effects of noise upon the direction of the vector $x_f \rightarrow x_g$.

In Figure 4.17, the error function $E_n(x_f)$ is evaluated for FOE-locations lying on a grid that is 10 pixels wide and covers an area of 200 by 200 pixels around the actual FOE. The size of the circle at each location indicates the amount of error, i.e., the deviation from the radial displacement field that would result if *that* location were picked as the FOE. Bold circles indicate error values which are above a certain threshold (5.0). Those FOE-locations that would result in displacement vectors which point *toward* the FOE (as described earlier) are marked as prohibited ($\oplus$). It can be seen that this two-dimensional error function is smooth and convex within a wide area around the actual FOE (marked by a small square). The shape of this error function makes it possible that, even with a poor initial guess for the location of the FOE, the global optimum can be found through local search methods.

Figures 4.18–4.23 show the effects of various conditions upon the behavior of this error function in the same 200×200 pixel square around the actual FOE as in Figure 4.17.

Effects of Displacement Vector Length

An important criterion is the function's behavior when the amount of camera translation is small or when the displacement vectors are noisy. Figure 4.18 shows how the shape of the error function depends upon the average *length* of the displacement vectors in the absence of any residual rotation or noise (except digitization noise). Clearly, the location of minimum error becomes more distinct with increasing amounts of displacement.

Effects of Residual Rotation

Figure 4.19 shows the effect of increasing residual *rotation* in horizontal direction upon the shape of the error distribution. Figure 4.20 shows the effect of residual rotation in *vertical* direction. Here, it is important to notice that the displacement field used is extremely non-symmetric along the Y-axis of the image plane. This is motivated by the fact that, in real ALV images, long displacement vectors are most likely to be found from points on the ground, which are located in the lower portion of the image. Therefore, positive and negative vertical rotations have been applied in Figure 4.20.

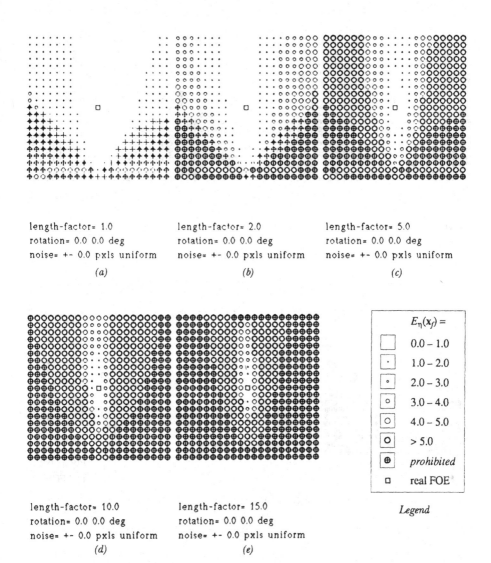

length-factor= 1.0
rotation= 0.0 0.0 deg
noise= +- 0.0 pxls uniform

(a)

length-factor= 2.0
rotation= 0.0 0.0 deg
noise= +- 0.0 pxls uniform

(b)

length-factor= 5.0
rotation= 0.0 0.0 deg
noise= +- 0.0 pxls uniform

(c)

length-factor= 10.0
rotation= 0.0 0.0 deg
noise= +- 0.0 pxls uniform

(d)

length-factor= 15.0
rotation= 0.0 0.0 deg
noise= +- 0.0 pxls uniform

(e)

$E_\eta(x_f) =$
0.0 – 1.0
1.0 – 2.0
2.0 – 3.0
3.0 – 4.0
4.0 – 5.0
> 5.0
prohibited
real FOE

Legend

Figure 4.18 Effects of average displacement vector length. The length factor varies from 1 to 15 (i.e, 4 to 60 pixels) in *(a-e)*. The error function (Equation 4.18) was evaluated over the same image area of 200×200 pixels around the actual FOE (small square) as in Figure 4.17. No rotation or noise was applied. Notice that the location of minimum error (and thus the location of the FOE) is better defined by longer displacement vectors.

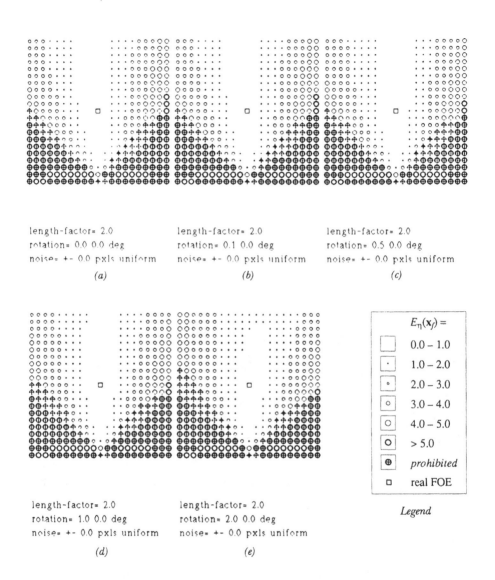

length-factor= 2.0
rotation= 0.0 0.0 deg
noise= +- 0.0 pxls uniform

(a)

length-factor= 2.0
rotation= 0.1 0.0 deg
noise= +- 0.0 pxls uniform

(b)

length-factor= 2.0
rotation= 0.5 0.0 deg
noise= +- 0.0 pxls uniform

(c)

length-factor= 2.0
rotation= 1.0 0.0 deg
noise= +- 0.0 pxls uniform

(d)

length-factor= 2.0
rotation= 2.0 0.0 deg
noise= +- 0.0 pxls uniform

(e)

$E_\eta(\mathbf{x}_f) =$

	0.0 – 1.0
.	1.0 – 2.0
∘	2.0 – 3.0
○	3.0 – 4.0
◯	4.0 – 5.0
⬤	> 5.0
⊕	*prohibited*
□	real FOE

Legend

Figure 4.19 Effects of horizontal camera rotations. The horizontal rotation varies from 0° to 2° in *(a–e)*. Displacement vectors are relatively short (length-factor = 2.0). No noise was applied.

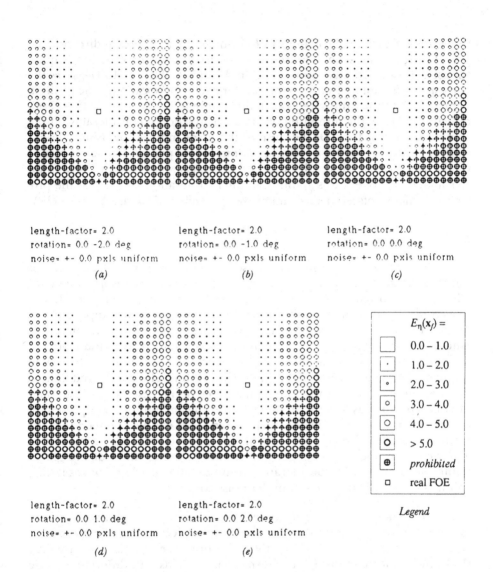

length-factor= 2.0
rotation= 0.0 -2.0 deg
noise= +- 0.0 pxls uniform

(a)

length-factor= 2.0
rotation= 0.0 -1.0 deg
noise= +- 0.0 pxls uniform

(b)

length-factor= 2.0
rotation= 0.0 0.0 deg
noise= +- 0.0 pxls uniform

(c)

length-factor= 2.0
rotation= 0.0 1.0 deg
noise= +- 0.0 pxls uniform

(d)

length-factor= 2.0
rotation= 0.0 2.0 deg
noise= +- 0.0 pxls uniform

(e)

	$E_\eta(\mathbf{x}_f) =$
□	0.0 – 1.0
·	1.0 – 2.0
∘	2.0 – 3.0
○	3.0 – 4.0
○	4.0 – 5.0
○	> 5.0
⊕	*prohibited*
□	real FOE

Legend

Figure 4.20 Effects of vertical camera rotations. The vertical rotation varies from –2° to 2° in *(a–e)*. The displacement vectors are relatively short (length-factor = 2.0). No noise was applied.

In Figure 4.21, residual rotations in *both* horizontal and vertical direction are present. It can be seen (Figures 4.21 *a–e*) that the error function is quite robust against rotational components in the image. Figures 4.21 *f–j* show the corresponding optimal shift vectors s_{opt} for each FOE-location. The result in Figure 4.21 *e* shows the effects of large combined rotations of 4° in both directions. Here, the minimum of the error function is considerably off the actual location of the FOE because of the error induced by using a linear shift to approximate the nonlinear derotation mapping. In such a case, it would be necessary to actually *derotate* the displacement field by the amount of rotation equivalent to s_{opt} found in the first iteration and repeat the process with the derotated displacement (see procedure *Evaluate–Single–FOE*).

Effects of Random Noise

The effects of various amounts of noise are shown in Figure 4.22. For this purpose, a random amount (with uniform distribution) of displacement was added to the original (continuous) image location and then rounded to integer pixel coordinates. Uniform random displacement was applied in the range of 0 to ±4 pixels in both the horizontal and the vertical direction. Since the displacement field contains only seven vectors, the results do not provide information about the statistical effects of image noise. This would require more extensive modeling and simulation. However, what can be observed here is that the absolute minimum error increases with the amount of noise. It can thus serve as an indicator for the amount of noise present in the image and the reliability of the final result. It can also be seen that the error distribution flattens out with increasing levels of noise, although its generic shape does not change. Furthermore, the location of the global minimum error may be considerably off the actual FOE which makes it difficult to locate the FOE precisely under noisy conditions.

Again, it should be noted that the length of the displacement vectors is an important factor. The shorter the displacement vectors are, the more difficult it is to locate the FOE correctly in the presence of noise. Figure 4.23 shows the error functions for two displacement fields with different average vector lengths. For the shorter displacement field (*length-factor* 2.0) in Figure 4.23 *a*, the shape of the error function changes dramatically under the same amount of noise (compare Figure 4.21 *a*). A search for the minimum error would inevitably converge towards a point (marked by the small arrow) that is far off the true FOE. For the image with *length-factor* 5.0 (Figure 4.23 *b*), the minimum of the error function coincides with the true location of the FOE (Figure 4.23 *a*).

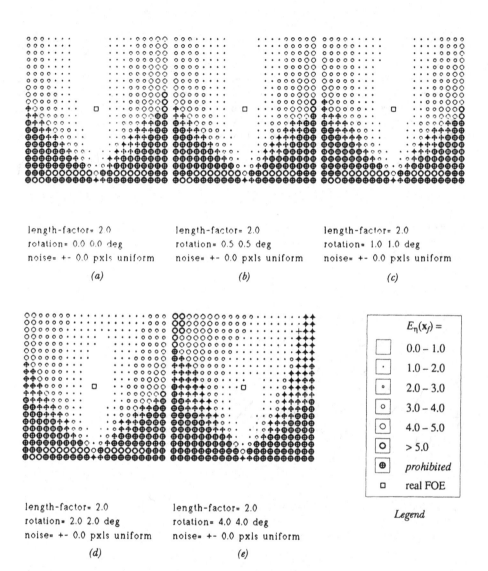

length-factor= 2.0
rotation= 0.0 0.0 deg
noise= +- 0.0 pxls uniform

(a)

length-factor= 2.0
rotation= 0.5 0.5 deg
noise= +- 0.0 pxls uniform

(b)

length-factor= 2.0
rotation= 1.0 1.0 deg
noise= +- 0.0 pxls uniform

(c)

length-factor= 2.0
rotation= 2.0 2.0 deg
noise= +- 0.0 pxls uniform

(d)

length-factor= 2.0
rotation= 4.0 4.0 deg
noise= +- 0.0 pxls uniform

(e)

$E_\eta(x_f) =$

☐	0.0 – 1.0
·	1.0 – 2.0
∘	2.0 – 3.0
o	3.0 – 4.0
○	4.0 – 5.0
◯	> 5.0
⊕	*prohibited*
☐	real FOE

Legend

Figure 4.21 Effects of combined horizontal and vertical camera rotations. The rotations (horizontal/ vertical) vary from (0°/0°) to (4°/4°) in *(a–e)*. The displacement vectors are relatively short (length-factor = 2.0). No noise was applied.

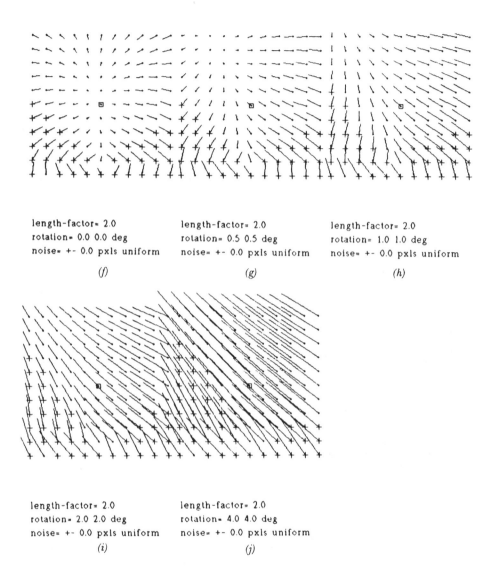

length-factor= 2.0   length-factor= 2.0   length-factor= 2.0
rotation= 0.0 0.0 deg   rotation= 0.5 0.5 deg   rotation= 1.0 1.0 deg
noise= +- 0.0 pxls uniform   noise= +- 0.0 pxls uniform   noise= +- 0.0 pxls uniform

           (f)   (g)   (h)

length-factor= 2.0   length-factor= 2.0
rotation= 2.0 2.0 deg   rotation= 4.0 4.0 deg
noise= +- 0.0 pxls uniform   noise= +- 0.0 pxls uniform
           (i)   (j)

Figure 4.21 (*continued*) The plots (*f–j*) show the optimal shift vectors s_{opt} for each FOE-location (Equations 4.14 and 4.16), corresponding to the error distributions in (*a–e*).

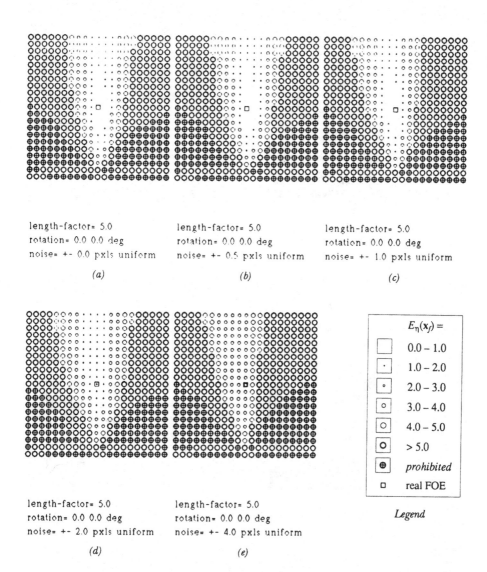

length-factor= 5.0
rotation= 0.0 0.0 deg
noise= +- 0.0 pxls uniform

(a)

length-factor= 5.0
rotation= 0.0 0.0 deg
noise= +- 0.5 pxls uniform

(b)

length-factor= 5.0
rotation= 0.0 0.0 deg
noise= +- 1.0 pxls uniform

(c)

length-factor= 5.0
rotation= 0.0 0.0 deg
noise= +- 2.0 pxls uniform

(d)

length-factor= 5.0
rotation= 0.0 0.0 deg
noise= +- 4.0 pxls uniform

(e)

$E_\eta(x_f) =$	
	0.0 – 1.0
·	1.0 – 2.0
∘	2.0 – 3.0
⊙	3.0 – 4.0
◯	4.0 – 5.0
⬤	> 5.0
⊕	*prohibited*
□	real FOE

Legend

Figure 4.22 Effects of random noise for constant average displacement vector
length. Uniform random noise was added to displacement vectors, ranging from 0
to ±4 pixels, in both directions *(a-e)*. Notice that, with increasing levels of noise, the
error distribution becomes increasingly flat around the location of minimum error.

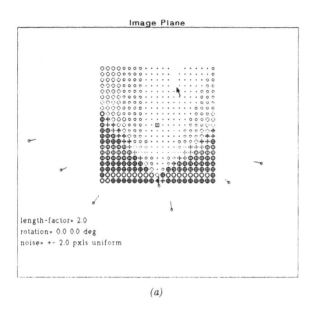

(a)

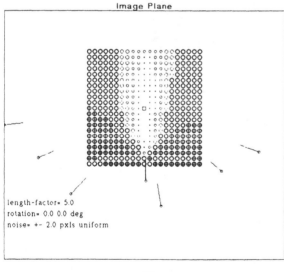

(b)

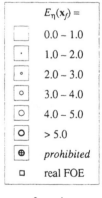

Legend

Figure 4.23 Effects of random noise for different average displacement vector lengths. For the short displacement field with length-factor 2.0 *(a)*, the disturbance moves the local minimum (arrow) far off the actual FOE. The same amount of noise applied to the longer displacement field with length factors 5.0 *(b)* has a much less dramatic effects.

Notice that, while the displacement field and the noise parameters are identical in Figures 4.22*d* and 4.23*b*, the resulting error distributions are different. This is because of different random values (noise) obtained in each experiment.

Conclusions

With the *Rotations-from-FOE* algorithm we have found a motion decomposition technique which works robustly under noisy conditions and which allows fast local search for the FOE. However, the last experiment (Figure 4.23) shows us two important things. *First*, we see that a sufficient amount of displacement (and thus camera translation) between successive frames is essential for reliably determining the FOE and therefore the direction of camera heading. *Secondly*, it appears that it may not be possible at all to locate the FOE precisely under realistic (i.e., noisy) conditions, using only a small number of displacement vectors. One could think of several remedies to this problem. For example, one could use much larger sets of displacement vectors, or make FOE measurements not just on two frames but on longer image sequences. A third alternative, which we will pursue in the next chapter, is to abandon the idea of a *precise* FOE entirely. Instead, we try to determine only the approximate motion parameters by using the *Rotations-from-FOE* approach and the properties of the spatial error distribution.

5

THE FUZZY FOE

As described in Chapter 4, the work towards the development of an FOE algorithm led to the following observations. The commonly proposed method (*FOE from Rotations*) for locating the FOE by intersecting the extended displacement vectors is inherently noise sensitive. The resulting error functions are not well behaved, making it difficult to find a global solution. In contrast, the second alternative discussed in Chapter 4 (*Rotations from FOE*) permits local search methods to be applied as a consequence of its error function being well behaved within a large region around the actual FOE. However, when the displacement vectors are short, noisy, or when there are only few vectors, the location of the error function's global minimum may not coincide with the actual FOE. In this case, the error function becomes increasingly flat and no distinct optimum exists.

5.1 AVOIDING UNREALISTIC PRECISION

Of course, it is possible to compute the location of the global minimum even in the case of an extremely flat error function. As a matter of fact, we could compute that location down to any precision we wanted, without getting any closer to the real FOE. In other words, the results may be *precise* but *not accurate.* Our goal is not only to avoid this kind of unrealistic precision but also to devise a mechanism for locating the FOE that accounts for the specific conditions that we discussed above.

Beyond the computational difficulties involved in locating the FOE, there are other motivations for reconsidering that concept. For example, humans also have difficulties when asked to point out the direction of heading under conditions of small translation (i.e., short displacement vectors) and/or large rotational components [60]. This, of course, does not imply that the human visual system performs an FOE computation closely to the way we do or even employ error functions related to ours. However, it may indicate that locating the FOE with high accuracy is indeed a very difficult task. But even with inaccurate knowledge about their own direction of heading, humans are able to make extremely good judgements about the 3-D layout and the movements of other objects in their environment. We can, there-fore, conclude that the precise knowledge of the FOE is not vital for per-forming some of the central tasks in motion understanding. In fact, the whole framework of qualitative reasoning about scene and image motion (covered in the Chapters 6 and 7) has been strongly influenced by this par-ticular observation. The "Fuzzy FOE", described in the following, is one of the two main components of this approach.

5.2   DEFINING THE FUZZY FOE

At this point we abandon the idea of locating the FOE as a single, distinct image location. Instead, we shall try to compute the direction of camera heading only to a precision that is related to the uncertainty in the underly-ing displacement data. In general, this means that the result will not be a single translation vector **T** (the 3-D camera translation) but a *set* of possible vectors. In terms of image coordinates, we are looking for a two-dimensional region of possible FOE locations for each successive pair of images. We have called such a region a "Fuzzy FOE" (FFOE). The two main questions that we are concerned with are: *(a)* what representation to use for the FFOE, and *(b)* how to compute it efficiently. The error function of (4.18) serves as the basis for computing the FFOE. As we have pointed out in the previous chap-ter, the local shape of this error function, i.e., its flatness, is an indicator for the quality of the underlying displacement field.

Whatever representation and means of computation we choose, we want the FFOE to have two essential properties. *First*, the definition (i.e., the spatial extent) of the FFOE should capture the amount of uncertainty contained in the original displacement data. For example, large camera translation should result in a small FFOE, while small translation or noise should pro-duce a larger FFOE. As the *second* requirement, we want to have good confi-dence that the FFOE includes the true FOE.

The FFOE itself could be represented in various different ways. For example, we could use a parametric description, such as a *circle* with origin x_0 and radius r. The problem of computing the FFOE would then be to find the size and position of a circle that best fits the error function $E_\eta(x_f)$ in (4.18). However, our experience has shown that the basins of low error values $E_\eta(x_f)$ are generally not circular but usually elongated, due to the non-uniform distribution of displacement vectors over the image. This may be a particular property of land vehicle image sequences, where most displacement vectors are found below the horizon (i.e., below the vertical position of the FOE). One could account for this fact by using, instead of a circle, an ellipsoid with a specific orientation and elongation to represent the FFOE. Of course, one could imagine even more sophisticated schemes that consider the fact that the real FOE is more likely to be located at the inner parts of the FFOE than at its periphery. For example, a two-dimensional Gaussian distribution could be a suitable model for the probability of FFOE locations.

We define the FFOE as a single compact region of potential FOE locations, i.e., a binary labeling of image points:

$$FFOE:\ \mathbb{N}_m \times \mathbb{N}_n \rightarrow \{0,1\} \tag{5.1}$$

Thus $FFOE(x,y) = 1$ *iff* the image point (x,y) is considered a potential FOE location. For practical reasons, a boundary representation of the FFOE is used instead of enumerating all interior image locations.

Compared to other alternatives mentioned above, the representation we have chosen may appear rather crude. However, for the purpose of the subsequent reasoning steps, the most important information supplied by the FFOE is the relative position and orientation of the displacement vectors. For example, we want to know if a displacement vector is clearly outside the FFOE, on which side it lies, or if two vectors are on opposite sides of the FFOE. While the boundary representation is sufficient for this purpose, the investigation of the more elaborate schemes may be worthwhile. Similarly, the following algorithm for computing the FFOE is very simplistic and could afford considerable improvements.

5.3   COMPUTING THE FUZZY FOE

The approach for computing the FFOE can be described as a 3-step process:

(1)   Guess a starting FOE, x_0.

(2)   From x_0, search for an image location x_{min}, such that the error $E_\eta(x_{min})$ is a minimum (e.g., following the steepest descent).

(3)   Around x_{min}, grow a connected region until some predefined criterion is met.

In particular, the region is complete when the accumulated *error volume* exceeds a predefined limit. Intuitively, the 2-D error function could be seen as a basin into which we pour a certain quantity of liquid (the error volume) and then take the boundary of the resulting surface as the outline of the Fuzzy FOE. The following algorithm *Compute-Fuzzy-FOE* and its auxiliary functions describe these steps in more detail. The function *Min-Neighbor* evaluates all image locations adjacent to a given region and returns the location with the smallest error value:

Compute-Fuzzy-FOE $(I_0, I_1, \text{error-lim})$:
   $x_0 := $ *Guess-Initial-FOE*;
   $x_{min} := $ *Search-Min-FOE* (x_0, I_0, I_1);
   ffoe $:= $ *Grow-Fuzzy-FOE* $(x_{min}, I_0, I_1, \text{error-lim})$;
   return (ffoe).

Search-Min-FOE (x_0, I_0, I_1):
   $x_{min} := x_0$;
   $(I^+, \phi_{min}, \theta_{min}, \text{err}_{min}) := $ *Evaluate-Single-FOE* (x_{min}, I_0, I_1);
   repeat
       $(x_n, \phi_n, \theta_n, \text{err}_n) := $ *Min-Neighbor* $(\{x_{min}\}, I_0, I_1)$;
       if $(\text{err}_n < \text{err}_{min})$ *then*
           $(x_{min}, \phi_{min}, \theta_{min}, \text{err}_{min}) := (x_n, \phi_n, \theta_n, \text{err}_n)$;
       until $(\text{err}_n \geq \text{err}_{min})$;
   return (x_{min}).

Grow–Fuzzy–FOE (x_{min}, I_0, I_1, error–lim);
   ffoe := $\{x_{min}\}$;   /* initialize the FOE region */
   err_{max} := err_{min};   /* max error found so far */
   error–vol := 0;
   repeat
      $(x_n, \phi_n, \theta_n, err_n)$:= *Min–Neighbor* (ffoe, I_0, I_1);
      ffoe := ffoe $\cup$ $\{x_n\}$;
      Δerr := err_n - err_{max};
      err_{max} := err_n;
      error–vol := error–vol + Δerr * *Size* (ffoe);
                    /* *Size* (ffoe) returns the area of the current FFOE-region */
   until (error–vol > error–lim);
   return (ffoe).

Min–Neighbor (Region, I_0, I_1):
   err_{min} := ∞;
   forall $x_i \in$ Region
     forall x_a: (*Neighbors–p* (x_a, x_i) & $x_a \notin$ Region)
      (I^+, ϕ_a, θ_a, err_a) := *Evaluate–Single–FOE* (x_a, I_0, I_1);
      if ($err_a < err_{min}$) *then*
      $(x_{min}, \phi_{min}, \theta_{min}, err_{min})$:= $(x_a, \phi_a, \theta_a, err_a)$;
     endfor
    endfor
   return $(x_{min}, \phi_{min}, \theta_{min}, err_{min})$.

An initial guess must be made to start the search down to the bottom of the error basin (procedure *Search–Min–FOE*). Due to the error function's wide range of convergence, the choice of the starting location is not critical. Figure 5.1 shows the traces of the search procedure for various starting locations. Also, since the direction of vehicle heading changes relatively slowly over time, the "best FOE" computed for the previous frame pair is usually a very close guess.

Once the bottom of the error basin has been localized, it becomes the seed for growing the FOE region. In procedure *Grow–Fuzzy–FOE*, the FOE region is iteratively dilated by adding in each step the neighboring location of minimum error. By selecting the element of smallest error adjacent to its current boundary, the FFOE is guaranteed to grow in a balanced fashion out of the center of the error basin. To compute the predicate *Neighbors–p*, any suitable neighborhood relation could be employed, such as the usual 8–neighborhood.

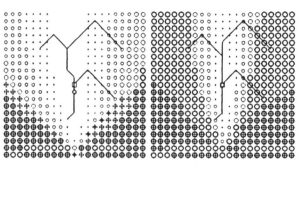

length-factor= 2.0
rotation= 0.0 0.0 deg
noise= +- 0.0 pxls uniform
(a)

length-factor= 5.0
rotation= 0.0 0.0 deg
noise= +- 0.0 pxls uniform
(b)

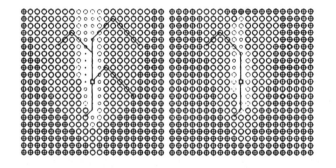

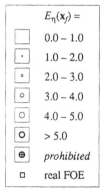

$E_\eta(\mathbf{x}_f) =$

□	0.0 – 1.0
·	1.0 – 2.0
∘	2.0 – 3.0
○	3.0 – 4.0
◯	4.0 – 5.0
◉	> 5.0
⊕	prohibited
□	real FOE

Legend

length-factor= 10.0
rotation= 0.0 0.0 deg
noise= +- 0.0 pxls uniform
(c)

length-factor= 15.0
rotation= 0.0 0.0 deg
noise= +- 0.0 pxls uniform
(d)

Figure 5.1 Searching for the bottom of the error basin. Traces of the search procedure **Search–Min–FOE** from various different starting locations are shown for displacement vector lengths 2, 5, 10, and 15 *(a–d)*. The real FOE is marked by a small square and the location of minimum error is marked by a small circle. The two locations are identical in *(b–d)*.

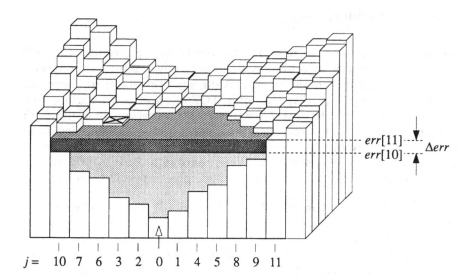

Figure 5.2 Updating the "error volume" when growing the FOE region. Each new image location $j = 0,1,...$ that is merged into the existing FOE region adds a slice to the current error volume whose thickness is proportional to the increase in error $\Delta err = err[j] - err[j-1]$. The drawing shows the situation after location $j = 11$ has been merged to the FOE region, such that a "slice" of thickness $\Delta err = err[11] - err[10]$ is added to the previous error volume. In the next step, the location of minimum error adjacent to the current FOE region would be merged (marked ✕). Location $j = 0$ marks the bottom of the error basin out of which the FFOE is grown.

In each iteration of *Grow–Fuzzy–FOE*, the total error is accumulated by computing the "error volume" (error–vol) contained in the basin covered by the current FOE region (Figure 5.2). Turning again to the analogy of a liquid being poured into the error basin, each new location adds a slice to the current error volume that has a certain thickness (Δerr) and the size of the resulting Fuzzy FOE region (*Size*(ffoe)). The growth terminates when the error volume reaches a predefined limit (error–lim) that was determined empirically.[1] Using the accumulated error volume to terminate growth means that flat error basins automatically lead to larger Fuzzy FOE regions than error functions with a sharp local minimum.

---

[1]   For error-lim, a value of 2500 was found to be appropriate. The limit values for small rotations ϕ_{max} and θ_{max} (used in procedure *Evaluate–Single–FOE*) were both set to 2°.

Practical Considerations

When it comes to implementing this approach there are various means for streamlining the basic algorithm. One obvious way to improve the speed of searching for the bottom of the error basin is to use a multi-grid approach, i.e., a variable step size. Given an initial FOE location as the starting point, the search along the steepest descent starts on a coarse grid at the beginning, continuously narrowing the grid size as it proceeds. The decision to switch to a finer resolution is made when no descent in any direction is possible with the current resolution. While there is, at least in principle, no lower limit to the grid size, the size of a single pixel is usually sufficient.

The function *Evaluate–Single–FOE* (see Section 4.3) is invoked by the above algorithm at several occasions, each time causing the computation of the optimal rotation angles and the amount of residual error for the hypothesized FOE location. The procedure *Search–Min–FOE* requires the repeated evaluation of point neighborhoods to proceed along the steepest path down into the error basin. In procedures *Grow–Fuzzy–FOE* and *Min–Neighbor* all locations adjacent to the current FOE region must be tested to determine the direction of growth. During the execution of either procedure, *Evaluate–Single–FOE* may be invoked on the same FOE-location several times. It is therefore useful to store the results of that function for the FOE-locations already visited.

A refined version of *Evaluate–Single–FOE* would first check if the result was already available in memory for the given location before actually running the computation and putting the result into memory. Since the FOE could potentially be located in any part of the image but only a small fraction of all image locations is ever evaluated, we have used a hash table that is indexed by the image location $(x\ y)$. Ultimately, given a suitable hardware, all hypothetical FOE locations could be evaluated in parallel because the individual computations are independent.

5.4 EXPERIMENTS

To evaluate the performance of the Fuzzy FOE algorithm under realistic conditions, it was applied to extended video sequences taken from the moving ALV. Figure 5.3 shows 16 frames of one of the original motion scenes.

Frame 182   Frame 183

Frame 184   Frame 185

Frame 186   Frame 187

Figure 5.3 Part of the original motion sequence taken from the moving ALV.

Frame 188   Frame 189

Frame 190   Frame 191

Frame 192   Frame 193

Figure 5.3 (*continued*).

Figure 194   Frame 195

Frame 196   Frame 197

Figure 5.3 (*continued*).

The spatial resolution of these images is 512×512 pixels, the vehicle traveled at approximately 14 kilometers per hour, and the elapsed time between each pair of frames is 0.5 seconds. Thus the 3-D translation vector had an average length of 1.9 meters.

Preprocessing

For these experiments, points were tracked manually in binary images that were obtained by edge enhancement and subsequent thresholding of the original images (Figure 5.3). Only the green color component of the original RGB image sequence was used. Edge enhancement was done with a compass operator that creates a set of 8 gradient images by convolving ($*$) each original image **G** with 8 directional templates. The maximum value at each image location is then taken as the gradient magnitude $\|\nabla\|$:

$$\|\nabla_k\| = \text{abs } (\mathbf{T}_k * \mathbf{G}) \qquad k = 0...7 \tag{5.2}$$

$$\|\nabla\| = \max_k \|\nabla_k\| \tag{5.3}$$

with

$$\mathbf{T}_0 = \begin{bmatrix} -1 & 0 & 1 \\ -\sqrt{2} & 0 & \sqrt{2} \\ -1 & 0 & 1 \end{bmatrix} \qquad \mathbf{T}_1 = \begin{bmatrix} 0 & \sqrt{2} & 1 \\ -\sqrt{2} & 0 & \sqrt{2} \\ -1 & -\sqrt{2} & 0 \end{bmatrix} \tag{5.4}$$

and similarly $\mathbf{T}_2...\mathbf{T}_7$, which are simply rotated versions of $\mathbf{T}_0$ and $\mathbf{T}_1$ respectively. The resulting gradient image $\|\nabla\|$ was then thresholded in the upper 20% of the histogram to obtain the binary edge image.

Token Tracking

By this simple kind of preprocessing, a considerable amount of detail is lost that usually helps humans to track image tokens, e.g., certain color cues. Experiments on both types of images showed that it is considerably more difficult to track points manually in the binary edge image than in the original grey-scale or color image. In highly textured areas of the image, the selection and tracking of reference points is almost impossible in the corresponding binary image. However, an algorithm could make use of the local structure in the original (color) images to track feature points reliably between frames. While we had no robust automatic tracking program available for these experiments, the assumption was that such a procedure is feasible and its performance can match manual tracking in binary images. Recent results [12, 38], in fact, strongly support the hope that fully automatic selection and tracking of feature points in real image sequences is indeed possible.

Figure 5.4 shows a sequence of 16 labeled frames obtained from the images in Figure 5.3. The reference point of each label is located at its lower left-hand corner. Approximately 25 tokens were selected in each image and the period of tracking varied from 2–16 frames.

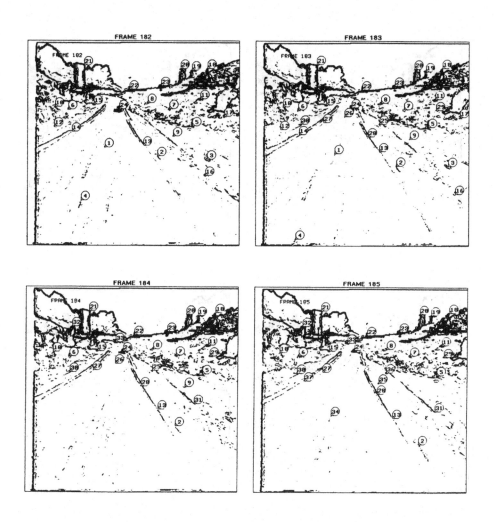

Figure 5.4 Binary ALV image sequence with selected tokens.

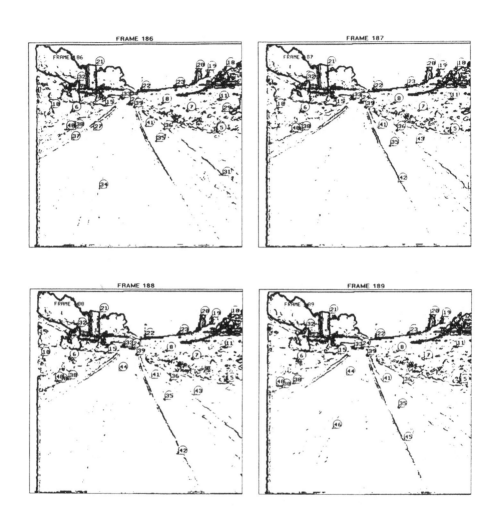

Figure 5.4 (*continued*).

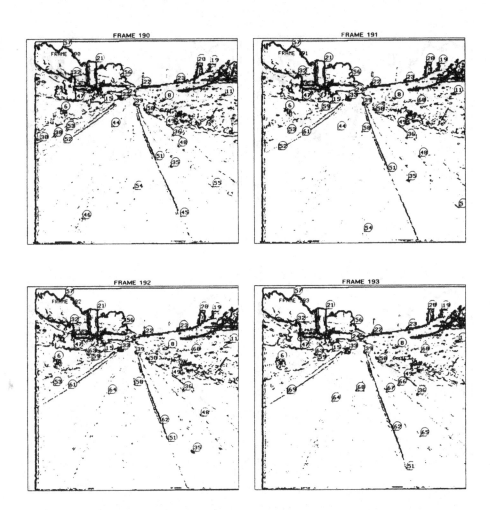

Figure 5.4 (*continued*).

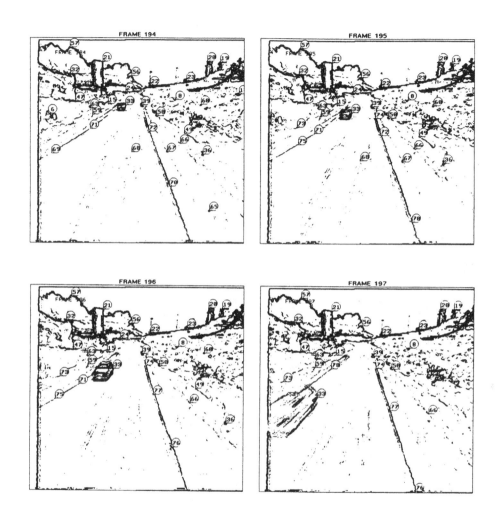

Figure 5.4 (*continued*).

For selecting tokens, we mainly considered local image features such as

- corners of lines (e.g., points 20 and 21 in Figure 5.4),
- end points of lines (e.g., point 31), and
- compact blobs (e.g., point 11).

Each token was assigned a unique label when encountered the first time. After the tracking of a point was started, its label remained unchanged until that point was no longer tracked. When no correspondence was found in the subsequent frame for a point being tracked, either because of occlusion, because the feature had left the field of view, or because it could not be identified any more, tracking of that token was discontinued. In case the same point reappeared again at a later time, it was treated as a new item and given a new label.

Tokens on the road surface (which are essential for estimating camera motion) turned out to be difficult to trace from frame to frame when they are fairly close to the camera. Since the edge enhancement operator applies the same mask regardless of the 3-D distance of the feature, the resulting binary image may change dramatically when features get close and "blow up" in the image. A successful automatic procedure for feature extraction and tracking in this kind imagery must account for this problem and will probably employ some form of range-dependent or multi-resolution technique [12].

Results

Figure 5.5 shows the results of applying the Fuzzy FOE algorithm to the image sequence of Figure 5.4. Each frame n displays the motion estimates for the period between n and the previous frame $n-1$. Therefore, the first motion estimate is available after the second frame ($n = 183$). In the search for the "best FOE", the optimal FOE-location x_0 (i.e., the one with minimum error $E_\eta(x_f)$) from the previous frame pair was taken as the initial guess. For the very first pair of frames, the FOE was guessed from the known camera orientation relative to the vehicle. Each motion estimate consists of three components: *(a)* the Fuzzy FOE, *(b)* the angles of the horizontal and vertical rotations, and *(c)* the approximate distance traveled over ground between each frame pair.

Fuzzy FOE Measurements

In Figure 5.5, the Fuzzy FOE is marked by a shaded area of varying shape and size. The jagged outline of the Fuzzy FOE is caused by the relatively coarse grid (10 pixels wide) used for growing the FOE region. The small circle inside this region marks the optimal FOE location. The shape of the FFOE depends strongly upon the most dominant (i.e., the longest) displacement vectors in the image. Since the most dominant vectors are usually found in the lower central parts of the image, the FFOE tends to be vertically elongated, i.e, the horizontal position of the FOE is usually better defined than its vertical position. This may be different, of course, for other types of scenes.

Rotation Estimates

The estimates for the rotations in horizontal and vertical direction are shown in a small coordinate frame with a range of ±1°. Due to the high weight of the ALV, the amount of rotation between frames is usually small and it was never necessary to apply an intermediate derotation during the FOE search. Along with the original displacement vectors (solid strokes), the vectors resulting from derotation are also shown in each frame (dotted strokes).

Absolute Vehicle Velocity

The absolute velocity of the vehicle is estimated after computing the FOE (see Appendix, Section A.2). The essential measure used for this computation is the *height* of the camera above the ground, which is approximately constant (3.3 meters). The estimated advancement (in meters) between each frame pair is shown in each frame of Figure 5.5. Notice that these numbers are only coarse approximations. For computing the absolute vehicle velocity, only a few prominent displacement vectors were selected in each frame pair. The criterion was that the vectors should be located below the FOE and their length should be more than 20 pixels. The end points of the selected (derotated) vectors are marked with dark dots.

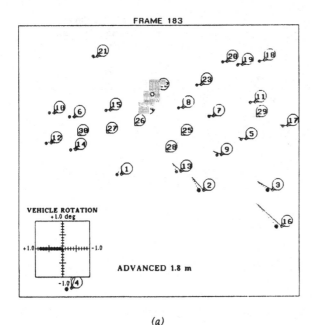

(a)

Figure 5.5 Results of the FOE computation. Numbered labels are located at the end points of the original displacement vectors. The shaded region is the Fuzzy FOE. The small circle inside the Fuzzy FOE marks the location of minimum error (i.e., the "best" FOE). The dotted lines are the derotated displacement vectors. The vectors that were used to compute the vehicle motion parameters are marked with heavy dots. The current estimate for the vehicle rotation is plotted in the lower left-hand corner of each frame. The absolute translation of the vehicle in Z-direction is measured in meters. Frames 183–184 (a–b).

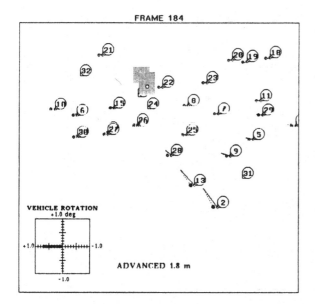

(b)

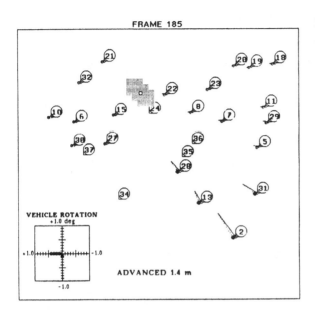

Figure 5.5 (*continued*).
Frames 185–186 (*c–d*).

(c)

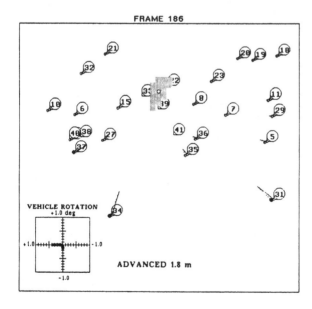

(d)

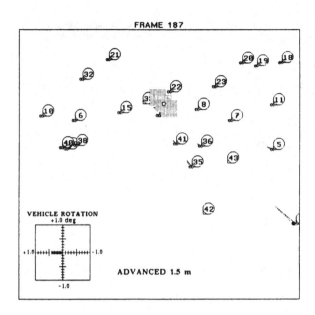

(e)

Figure 5.5 (*continued*).
Frames 187–188 (*e–f*).

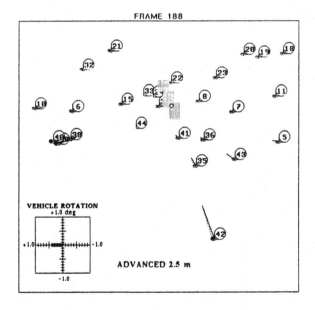

(f)

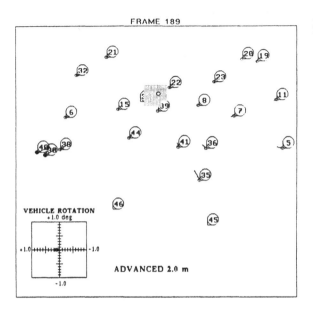

(g)

Figure 5.5 (*continued*).
Frames 189–190 (*g–h*).

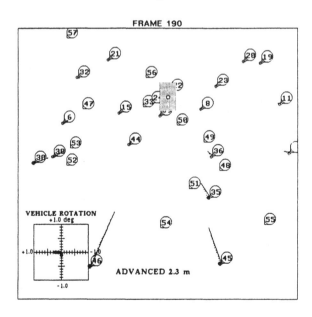

(h)

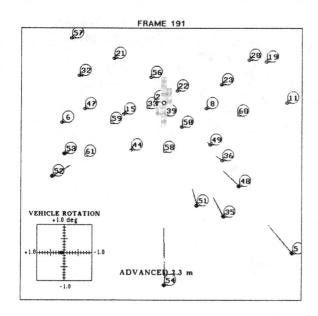

(i)

Figure 5.5 (*continued*).
Frames 191–192 (*i–j*).

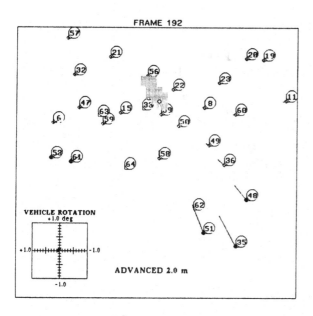

(j)

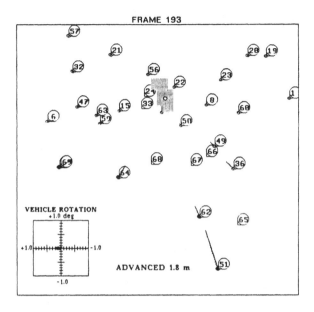

Figure 5.5 (*continued*).
Frames 193–194 (*k–l*).

(k)

(l)

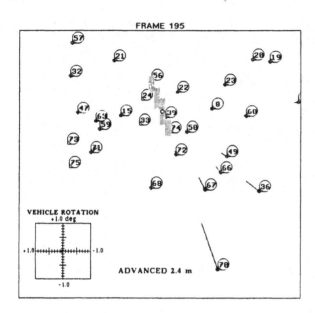

(m)

Figure 5.5 (*continued*).
Frames 195–196 (*m–n*).

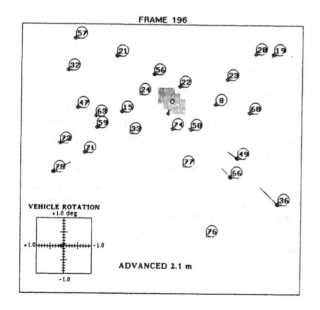

(n)

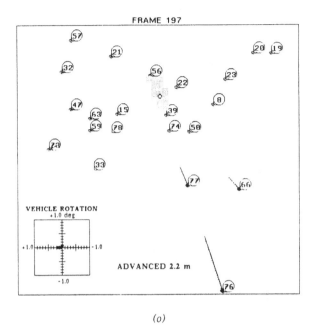

Figure 5.5 (*continued*).
Frame 197 (*o*).

(*o*)

Use of the Fuzzy FOE

The Fuzzy FOE as a tool for computing the camera's motion parameters plays a central role in our overall approach to motion analysis (see Figure 2.2). In the following chapters, we use the Fuzzy FOE and the properties of the derotated displacement vectors to derive information about the spatial layout of the scene and to detect the motion of other objects. In all the reasoning steps involved, we take into account that we do not know the FOE (and the camera rotations) accurately but only in approximate terms. As it will become evident, several interesting inferences about 3-D structure and motion can be made even *without* knowing the exact location of the FOE. Of course, we cannot expect these inferences to be more accurate than the underlying data. In particular, if the Fuzzy FOE is small (i.e., well defined), the information derived from it may also be more accurate than with a large Fuzzy FOE.

6

REASONING ABOUT
STRUCTURE AND MOTION

The material presented in the preceding chapters (Chapters 3–5) was mainly concerned with computing the camera's motion parameters from the composite displacement field. The result of this computation is the direction of camera heading (specified by the Fuzzy FOE), the angles of camera rotation about the horizontal and vertical axes, and the *derotated* displacement field for each pair of frames. As pointed out in Chapter 2 (Figure 2.2), this is the initial stage of the "DRIVE" approach. In this chapter and the two subsequent chapters, we describe the higher levels of the DRIVE approach, in particular the *3-D Motion Interpretation* stage and the *Qualitative Scene Model.*

The purpose of the higher-level DRIVE components (Figure 6.1) is to interpret the derotated displacement field to obtain information about the 3-D scene structure and individual object motion. The derotated displacement field is obtained from the original displacement field by removing the effects of the estimated camera rotations. It thus consists only of the displacement caused by the camera translation and possibly the displacement caused by other moving objects in the scene.

In the *3-D Motion Interpretation* process, certain interesting events are selected from the derotated displacement field and coded into symbolic descriptions. For this purpose, the configurations of the displacement vectors relative to each other and relative to the Fuzzy FOE are observed. Two sets of rules use the selected events to create and continually update a 3-D description of the scene, called the *Qualitative Scene Model* (QSM). The *generation rules* search for salient image events that can be used to draw direct conclusions about the 3-D reality. For example, a displacement vector pointing *toward* the FOE is a clear indicator for object motion (assuming the camera is moving forward). In contrast, the *verification rules* check existing hypotheses (in the QSM) for consistency with the observed image events.

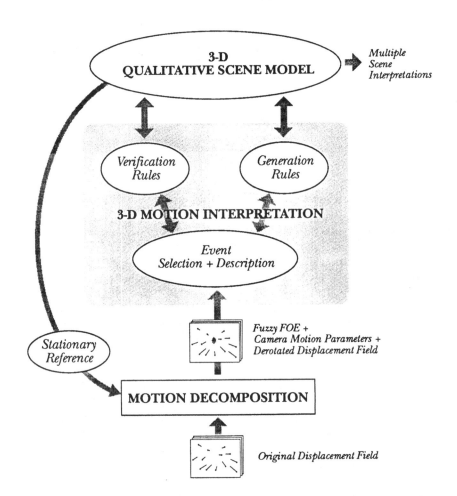

Figure 6.1 Structure of the 3-D interpretation process. First, significant image events are extracted from the derotated displacement field and coded into symbolic descriptions. Two sets of rules operate on these abstracted image events: *Generation rules* scan the set of events in a bottom-up fashion and place immediate 3-D conclusions (hypotheses) in the scene model. *Verification rules* check existing hypotheses for consistency with the changes occurring in the displacement field. The location of the Fuzzy FOE is used by most of the rules. A set of environmental entities that are believed to be stationary is supplied by the scene model to be used as a reference in the FOE computation (*Stationary Reference*).

The use of the Fuzzy FOE (instead of an accurate FOE) is directly reflected in the coding of image events as well as in the design of the inference rules. In fact, our qualitative reasoning approach itself was originally motivated by the inexact nature of the Fuzzy FOE and appears to us as a natural extension. The remaining parts of this chapter focus on the description of image events and the design of the rule base. In Chapter 7, we outline the internal structure of the *Qualitative Scene Model* and its dynamic behavior, followed by some illustrative experiments in Chapter 8.

6.1 ABSTRACTING IMAGE EVENTS

Disregarding the effects of spatial sampling, the changes that occur in the image of a moving camera are continuous. In order to apply some form of symbolic reasoning in this domain, we must first extract symbolic information from the continuous stream of input data. These symbolic tokens may either convey static information (e.g., the image position of a point with respect to the FOE) or dynamic information (e.g., the relative movement of two image points). Regardless of their nature, we refer to these symbolic pieces of data as *image events*. They are effectively abstractions of specific configurations and changes that are apparent in the image sequence.

Some of the image events are detected bottom-up, i.e., independent of the current state of the reasoning process. These are particularly salient events (termed *trigger events*) that may directly lead to conclusions about the 3-D scene. Other, less salient image events are created on demand in a top-down fashion. In the following, we describe the semantics of the most prominent image events used in our system. While some events have a very simple meaning, others call for a more subtle description.

Static 2-D Relations

Locations of features *relative to each other* (in the image):

- (LEFT-OF a b t), (RIGHT-OF a b t):
    Feature a is to the left (right) of b at time t;

- (ABOVE a b t), (BELOW a b t):
    Feature a is above (below) b at time t;

Locations of features *relative to the FOE*:

- (LEFT-OF-FOE a t), (RIGHT-OF-FOE a t):
    Feature a is to the left (right) of the FOE at time t;

- (ABOVE-FOE a t), (BELOW-FOE a t):
    Feature a is above (below) the FOE at time t;

    Since, in general, not a single FOE location is given, the above relationships must be interpreted with respect to a *set* of possible FOE-locations (i.e., the Fuzzy FOE). For example, (LEFT-OF-FOE a t) is only true when a is left of *every* possible FOE-location $\mathbf{x}_f$ at time t:

    $$\text{(LEFT-OF-FOE } a\ t) \Leftrightarrow \{x_a < x_f \ \text{ for all } \mathbf{x}_f = (x_f\,y_f) \in FFOE(t)\}.$$

Two particularly useful relationships that derive from the ones above are OPPOSITE-TO-FOE and INSIDE:

- (OPPOSITE-TO-FOE a b t):
    Features a and b lie on opposite sides of the FOE at time t, which is equivalent to

    (LEFT-OF-FOE a t) & (RIGHT-OF-FOE b t) or

    (ABOVE-FOE a t) & (BELOW-FOE b t).

- (INSIDE a b t):
    Feature a is closer (in 2-D) to the FOE than b. Again, this is measured relative to the set of possible FOE-locations:

    $$(\text{INSIDE } a\ b\ t) \Leftrightarrow \tag{6.1}$$

    $$\{d(\mathbf{x}_f,\mathbf{x}_a) < d(\mathbf{x}_f,\mathbf{x}_b) \quad \text{for all } \mathbf{x}_f \in FFOE(t)\} \Leftrightarrow$$

    $$\{d(\mathbf{x}_f^{b,min},\mathbf{x}_a) < d(\mathbf{x}_f^{b,min},\mathbf{x}_b) \ \text{ and } \ d(\mathbf{x}_f^{a,min},\mathbf{x}_a) < d(\mathbf{x}_f^{a,min},\mathbf{x}_b)\}$$

    where $\mathbf{x}_f^{b,min}$ is the FFOE-location closest to $\mathbf{x}_b$ and $d()$ is the Euclidean distance in 2-D:

$$d(\mathbf{x}_f^{b,min},\mathbf{x}_b) < d(\mathbf{x}_f^i,\mathbf{x}_b) \quad for\ all\ \mathbf{x}_f^i \in FFOE(t) \tag{6.2}$$

If two features are close together in the image, the INSIDE-relationship with respect to the FOE area should be easy to determine (Figure 6.2). For two features lying in different parts of the image, the INSIDE-relationship can only be established when one feature is clearly closer to any possible FOE-location than the other feature. The above formulation takes this into account without explicitly distinguishing between these two cases.

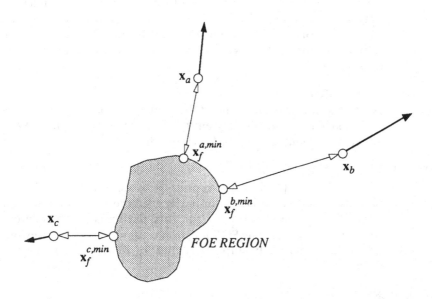

Figure 6.2 Illustrating the INSIDE-relationship. Image features *a, b, c* and the Fuzzy FOE (shaded area) are given. $\mathbf{x}_f^{a,min}$ is the FOE-location which is closest to $\mathbf{x}_a$, $\mathbf{x}_f^{b,min}$ is the FOE-location which is closest to $\mathbf{x}_b$. Here (INSIDE *a b t*) is true, but neither (INSIDE *a c t*) nor (INSIDE *b c t*).

Dynamic Image Events

Dynamic image events express the salient changes that occur in the image over time.

- (NEW-FEATURE *a t*):
> Feature *a* is observed for the first time at time *t*. Whenever this fact is asserted, a new entity *a* is added to the model and hypothesized to be stationary (by default).

- (LOST-FEATURE *a t*):
> Tracking of feature *a* was discontinued at time *t*. When a feature is lost by the correspondence process, the resulting conclusions in the model depend on the location of this feature in the image at time *t*–1. If the feature was close to the border of the image at time *t*–1, it has probably left the field of view at time *t*. In that case, the corresponding entity is removed from the model. Otherwise, the feature was lost due to changing viewing conditions or became *occluded*.

- (MOVING-TOWARDS-FOE *a t*):
> Feature *a* moves towards the FOE at time *t*. This is a strong indicator that the corresponding entity *a* is actually moving in 3-D and affects the *Qualitative Scene Model* immediately.

- (DIVERGING-NONRADIAL *a t*):
> Feature *a* does not diverge from the FOE along a straight line at time *t*. This situation is manifested by a substantial deviation from a radial expansion pattern (for this particular displacement vector) at any possible FOE-location. Along with the previous observation, this is an important clue for detecting 3-D motion directly.

- (CONVERGING *a b t*):
> Two features *a* and *b* are getting closer to each other in the image. This does not imply that any of them is actually moving in 3-D. The conclusions drawn from this fact depend on the particular locations of the two features relative to the FOE. This relationship is symmetric:

$$\text{(CONVERGING } a\ b\ t) \Leftrightarrow \text{(CONVERGING } b\ a\ t).$$

• (DIVERGING-FASTER *a b t*):

Feature *a* appears to be diverging faster from the FOE than *b*. Again, the rate of divergence is measured with respect to the entire *set* of possible FOE-locations. Given the image location x_a of *a* at time *t* and x_a' of *a* at time *t'* the divergence[1] of x_a with respect to a single FOE x_f is given by

$$div(x_a, x_f) = \frac{d_a}{r_a} = \frac{[x_a - x_f] \cdot [x_a' - x_a]}{\|x_a - x_f\|^2} \qquad (6.3)$$

where · denotes the inner product. Figure 6.3 shows the geometric interpretation of (6.3). In order to establish (DIVERGING-FASTER *a b t*) with respect to a *set* of possible FOE-locations *FOE(t)* it must be guaranteed that

$$div(x_a, x_f) > div(x_b, x_f) \qquad \text{for all } x_f \in FOE(t). \qquad (6.4)$$

To avoid checking this relation at each possible FOE-location, it is sufficient to verify that both

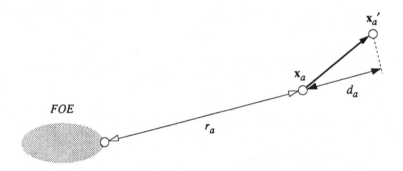

Figure 6.3 Measuring the rate of divergence (*div*). d_a is the length of the projection of the displacement vector $x_a' - x_a$ onto the extended vector $x_f - x_a$.

---

[1]   Not to be confused with the term *divergence* in the sense of vector analysis.

$$div(\mathbf{x}_a, \mathbf{x}_f^{a,max}) > div(\mathbf{x}_b, \mathbf{x}_f^{a,max}) \quad \text{and}$$
$$div(\mathbf{x}_a, \mathbf{x}_f^{b,min}) > div(\mathbf{x}_b, \mathbf{x}_f^{b,min}) \tag{6.5}$$

are satisfied. Here $\mathbf{x}_f^{a,max}$ is the FOE-location which is farthest from $\mathbf{x}_a$ and $\mathbf{x}_f^{b,min}$ is the FOE-location which is closest to $\mathbf{x}_b$. These particular FOE-locations are supplied for every feature when the FOE is actually computed.

- (PASSING a b t):

    Feature a is *inside* b and a is getting closer to b. Since the *passing* entity is closer in 3-D (assuming that both a and b are stationary), this is an important clue for determining the relative 3-D positions of entities a and b, also called the *deletion effect*. Figure 6.4 shows a situation where one feature is passing another in the image.

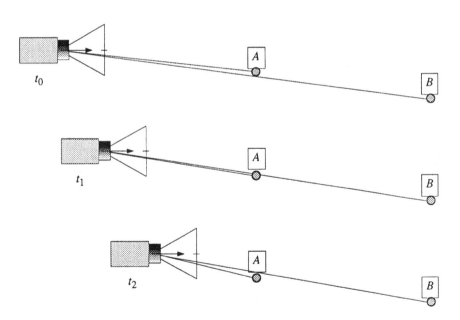

Figure 6.4 Illustrating the PASSING-relationship. The camera (top view) is moving from left to right in a scene that contains two stationary objects A and B. Object A is closer to the camera than B and appears to be left of B at time t_0. As the camera is moving forward, the images of A and B converge (t_1) and eventually A is observed to the *right* of B (t_2). Thus A has "passed" B from the inside to the outside.

At the heart of our motion analysis scheme is the *Qualitative Scene Model* that will be described in Chapter 7. The facts contained in this model comprise the system's knowledge about the 3-D scene at any point in time. All the image events being created in the bottom-up analysis of the displacement fields are entered as corresponding facts into model. Image tokens and image frames are assigned unique numbers. For example, the event (CONVERGING *a b t*) for *a* = 36, *b* = 17, *t* = 183, would cause

(CONVERGING 36 17 183)

to be added to the set of known facts in the scene model. The facts in this global data base are continually observed by a set of rules that, upon firing, may add new facts and remove existing facts. For example, the PASSING relationship is asserted by the following rule:[1]

```
❑    (defrule DETERMINE-PASSES
       (INSIDE ?x ?y ?t)
       (not (OPPOSITE-TO-FOE ?x ?y ?t))
       (CONVERGING ?x ?y ?t)
       =>
       (assert (PASSING ?x ?y ?t))) .
```

In this example, a new fact (PASSING *x y t*) is asserted (i.e., added to the global fact base) when all three preconditions of the rules are satisfied. While the rule itself is written in a forward-chained (i.e., bottom-up) manner, this does not imply that the preconditioning facts are obtained bottom-up as well. In ART, the rule antecedents are evaluated in sequential order, i.e., the condition (INSIDE...) would be checked first and (CONVERGING...) last. If a rule for asserting the fact (CONVERGING...) were backward-chained, it would not be fired until there is a need to check that relation for *specific* arguments.

---

[1] We use the original syntax of ART (a trademark of Inference Corp.) for defining rules which is both readable and concise. The (simplified) format of a rule definition is
```
(defrule rule-name
    condition-1 condition-2   … condition-m
    =>
    action-1 action-2 … action-n) .
```

6.2 INTERPRETING IMAGE EVENTS

Image events, correspond to the observations made in the 2-D image but do not convey explicit information about the 3-D reality. However, scene interpretations are formed by a reasoning process which essentially relies upon the image events described above. This reasoning process can be thought of a set of independent knowledge sources (rules) that operate around a common set of data (facts).

Each *token* observed in the image sequence has a counterpart in the 3-D scene model, called an *entity*. All the properties and relationships defined in the following apply to entities rather than tokens. Entities, their properties and relationships are the building blocks of the 3-D scene interpretations. Many of the expressions dealing with scene entities do not represent known facts but are merely temporal hypotheses that may be withdrawn at some future point. The differences between facts and hypothesis with regard to the model structure will be explained in more detail in Chapter 7.

In the following, we describe the two main sets of knowledge sources that contribute to the construction of scene interpretations. The first group is concerned with the *stationary* part of the environment that is being traversed by the vehicle. The second group of rules aims at detecting and classifying the 3-D motion of objects in the environment.

A scene interpretation consists of a set of *stationary* objects and a (possibly empty) set of *mobile* objects. Each entity A within an interpretation is either considered stationary or mobile, expressed by

- (STATIONARY A) or (MOBILE A),

respectively. In order to be consistent, an interpretation may not contain an entity that is considered both stationary *and* mobile. When an entity is observed for the first time, it is assumed to be stationary by default. If an entity has been found to be in motion once, it is considered *mobile* forever from that point on. However, before looking at the details of 3-D motion description, we first focus on the stationary entities of the scene (or rather those that are thought to be stationary).

6.3 REASONING ABOUT 3-D SCENE STRUCTURE

Modeling Scene Structure

The model that we use is a viewer-centered representation in which the locations of features in the image specify two dimensions. The third dimension is the depth of entities, i.e., their distance (in Z-direction) from the camera plane. The advantage of a viewer-centered representation is that there is only a single coordinate system. When the camera is moving, however, the distances of points in the scene change continually. A regular (i.e., metric) geometric model of the environment would therefore require a complete update after each frame. Another problem is that, due to the uncertainty involved in computing the FOE, the absolute depth of scene entities can only be estimated but not determined exactly. A useful model must account for this uncertainty.

Consequently, measuring absolute depth is not a major concern in our modeling scheme. Instead, we construct a *partial ordering* of scene entities with respect to their distance from the image plane. This qualitative depth map does not require updating after each frame but is continually refined over multiple frames as more evidence about the scene layout becomes available. Central to this concept is the CLOSER relationship.

- (CLOSER A B):
    Entity A is believed to be *closer* to the image plane than entity B. This relationship only applies to stationary entities. CLOSER is transitive, i.e.,
    (CLOSER A B) & (CLOSER B C) $\Rightarrow$ (CLOSER A C).

Figure 6.5 shows an example how a network of *closer*-relationships is constructed in the case of a completely stationary set of features A, B, C, D. Initially, nothing is known about their spatial relationships except that, since they are visible, all are in front of the image plane. This situation is reflected by the graph in Figure 6.5a. As the camera moves forward, new CLOSER-pairs are added to the graph, thus creating a partial ordering of scene entities with respect to their depth in 3-D space.

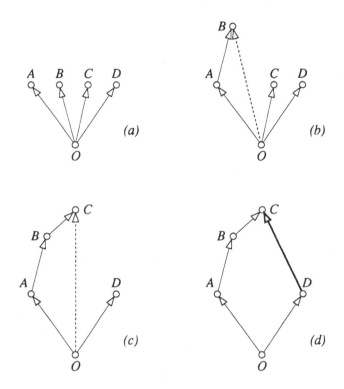

Figure 6.5 Successive construction of a partial ordering in depth. *(a)* Initial situation: no relative depth is known for the scene entities *A, B, C, D*. They are only known to be in front of the image plane *O*. *(b)* (CLOSER *A B*) has been determined and is added to the set of facts. *(c)* (CLOSER *B C*) has been determined, (CLOSER *A C*) is implied by transitivity. *(d)* (CLOSER *D C*) has been determined. Notice that, at this point, nothing can be said about the spatial relationships between feature pairs *(A,D)* and *(B,D)*.

Inferring "CLOSER"

The primary observation that leads to the assertion of the CLOSER-relationship is the relative rate of divergence of image features with respect to the FOE. Due to the uncertainty expressed by the Fuzzy FOE we cannot measure the rate of divergence accurately for every image token (see Equation

6.4). However, the fact (DIVERGING-FASTER *a b t*), if true, provides strong evidence that the entity *A* (whose image is *a*) is indeed closer than *B* in 3-D space. Of course, both entities in a CLOSER pair must be considered *stationary* in the corresponding scene interpretation, as reflected by the following rule:[1]

❑   **(defrule** CLOSER-FROM-EXPANSION
    (CURRENT-TIME ?t)
    (STATIONARY ?A)
    (STATIONARY ?B)
    (DIVERGING-FASTER ?a ?b ?t)
    =>
    (assert (CLOSER ?A ?B)))

Since this rule applies equally to any pair of image tokens, it allows to find CLOSER pairs that span the entire field of view. This makes it possible to relate the depths of entities that lie on opposite sides of the FOE, e.g., objects located to the left and to right of the road. However, locally different rates of divergence, implied by the PASSING relationship can supply even stronger evidence:

❑   **(defrule** CLOSER-FROM-PASSES
    (CURRENT-TIME ?t)
    (PASSING ?a ?b ?t)
    (STATIONARY ?A)
    (STATIONARY ?B)
    =>
    (assert (CLOSER ?A ?B)))

The reason for including this second rule is that the FOE computation may not return a perfectly *derotated* (i.e., radial) displacement field or that the Fuzzy FOE can be very large. In such a case, it may be difficult to measure the individual rates of expansion. However, *passes* between features can be detected even in the presence of considerable residual rotation or when only a very rough estimate of the FOE is known. Consequently, the rule CLOSER-FROM-PASSES is more selective but also more reliable.

---

[1]   Variable names in ART rules are of the form *?name*. We use upper and lower case variable names here only for the purpose of clarity. Lower case names like ?a refer to image tokens, while uppercase names like ?A denote 3-D entities. The rule language itself is not case sensitive, i.e., ?A is equivalent to ?a.

The validity of the rule CLOSER–FROM–PASSES for two features a and b follows from (3.20):

$$Z_a = \Delta Z \frac{\|\mathbf{x}_a' - \mathbf{x}_f\|}{\|\mathbf{x}_a' - \mathbf{x}_a\|} = \Delta Z \frac{r_a'}{d_a} = \Delta Z \left(1 + \frac{r_a}{d_a}\right) \tag{6.3a}$$

$$Z_b = \Delta Z \frac{\|\mathbf{x}_b' - \mathbf{x}_f\|}{\|\mathbf{x}_b' - \mathbf{x}_b\|} = \Delta Z \frac{r_b'}{d_b} = \Delta Z \left(1 + \frac{r_b}{d_b}\right) \tag{6.3b}$$

The conditions that must hold in order to assert (PASSING $a\ b\ t$) are:

$$(\text{INSIDE } a\ b\ t) \wedge \neg (\text{OPPOSITE } a\ b\ t) \wedge (\text{CONVERGING } a\ b\ t)$$

where

$$(\text{INSIDE } a\ b\ t) \Rightarrow r_a < r_b$$

and

$$(\text{CONVERGING } a\ b\ t) \Rightarrow d_a > d_b.$$

Since $r_a, r_b, d_a > 0$ and $d_b \geq 0$, it follows that

$$\frac{r_a}{d_a} < \frac{r_b}{d_b} \ \Rightarrow \ Z_a = \Delta Z \left(1 + \frac{r_a}{d_a}\right) < \Delta Z \left(1 + \frac{r_b}{d_b}\right) = Z_b \tag{6.4}$$

So, if (PASSING $a\ b\ t$) is true, then Z_a is less than Z_b and therefore A is actually closer to the camera plane than B in 3-D (provided that both A and B are stationary, of course).

6.4   REASONING ABOUT 3-D MOTION

Modeling 3-D Motion

Similar to the way we describe the spatial relationships of stationary entities, the motion of objects in 3-D space is also represented in coarse, qualitative terms. Depending upon the evidence that is available to the system from image events, the description of objects motion is more or less specific.

The least specific statement that can be made about a moving entity *A* is (MOBILE *A*). In order to be considered *mobile*, an entity does not have to be actually in motion. It is sufficient if the entity has been observed in motion *once* to remove it from the set of stationary objects of a scene interpretation. If an entity is currently observed moving in 3-D, this is reflected by the predicate

- (MOVES *A t*)

Of course, (MOVES *A t*) implies (MOBILE *A*). If a particular direction of motion (with respect to the camera-centered coordinate system) has been determined, this is expressed by

- (MOVES-LEFT *A t*)	- (MOVES-RIGHT *A t*)
- (MOVES-UP *A t*)	- (MOVES-DOWN *A t*)
- (MOVES-APPROACHING *A t*)	- (MOVES-RECEDING *A t*).

Inferring 3-D Motion

The detection of 3-D motion in the scene is accomplished in two ways.

- *Direct* motion detection is based on clues in the image, which directly imply 3-D motion and do not require any knowledge about the structure of the scene.

- *Indirect* motion detection interprets the changes in the image in the light of the hypothesized 3-D scene structure. Inconsistencies between those changes and the current contents of the model generally lead to new scene interpretations, in which additional entities are considered mobile.

Direct Motion Detection

For a stationary scene, every point in the derotated image should be diverging away from the FOE. A displacement vector pointing *towards* the FOE cannot be caused by a stationary entity. A similar but weaker condition for detecting motion directly is when a feature does not diverge *radially* out of the FOE. This is indicated by a large individual error for the particular displacement vector in the course of the FOE-computation. Unless the FOE computation had gone entirely wrong, either observation would clearly indicate some form of 3-D motion. Both conditions are considered by the following rule:

❑   **(defrule** DIRECT-SINGLE-MOTION
           (or   (MOVING-TOWARDS-FOE ?x ?t)
                     (DIVERGING-NON-RADIAL ?x ?t))
           =>
           (at ROOT (assert (MOBILE ?X)))) .

The directive "at ROOT" in the action part of this rule has a special meaning that will be explained in Chapter 7. In short, it causes the new fact (MOBILE ?X) to be asserted at the root node of the interpretation graph. By doing so, the fact becomes *global* and thus visible in every existing hypothesis. This possibly creates local conflicts. For example, if ?X is bound to the value A, then any hypothesis that contains (STATIONARY A) would be *removed* because of the resulting inconsistency (see below).

The *second* category of direct motion detection involves *pairs* of features and does not require a perfectly derotated image like the previous rule. It makes use of the fact that stationary features lying on *opposite* sides of the FOE may not converge. Therefore, only a rough knowledge of the FOE-location must be available to apply this rule:

❑   **(defrule** DIRECT-PAIR-MOTION
           (OPPOSITE-TO-FOE ?x ?y ?t)
           (CONVERGING ?x ?y ?t)
           =>
           (at ROOT (assert (MOVEMENT-BETWEEN ?A ?B)))).

After firing this rule, the system assumes that there is relative motion between two particular entities in 3-D. Again, this is not taken as a hypothesis but as a known (i.e., global) fact. At this point it is not concluded *which* of the two entities are in motion. This part of the reasoning chain is discussed in Chapter 7.

Indirect Motion Detection

While some forms of 3-D object motion are easily detected, other modes of motion will pass unnoticed by the above inference rules. The main reason for this is that moving objects may *pretend* to be stationary in 3-D by producing an image trace that is compatible with the hypothesis of a non-moving object. Of course, this does not suggest that objects can usually control their motion in order to appear stationary to some moving observer. However, certain configurations can produce this effect in real-world scenes.

As an example, consider the situation in Figure 6.6, which shows a side-view of the vehicle. The vehicle (and the camera) is moving at a constant speed,

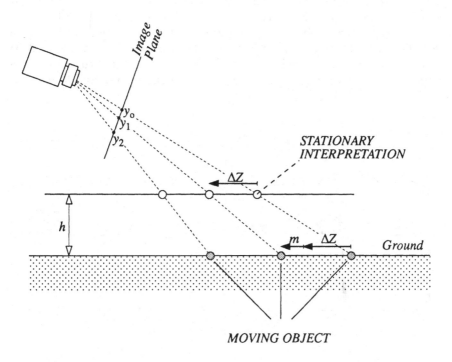

Figure 6.6 Moving objects may appear to be stationary. An object on the ground that is moving towards the camera at a rate of m per frame is observed in three frames. The camera moves forward at a rate of ΔZ per frame. If m is constant, a stationary interpretation, with the object located at height h in 3-D, is perfectly legitimate. The true movement would go undetected by purely geometric reasoning.

advancing by a distance of ΔZ between each pair of frames. A single feature is observed in three frames, at image locations y_0, y_1, y_2. This feature is assumed to be stationary, but in reality it belongs to an object that is moving towards the camera with a velocity of m units per frame. Assuming that m is constant within the period of observation, it would be legitimate to consider that entity to be *stationary* at height h above the ground, although it is actually moving in 3-D.

This means that some forms of continuous motion cannot be detected by means of a purely *geometric* interpretation. In particular, objects may not be detected as moving when they are actually approaching the camera on a straight path with constant velocity. Figure 6.7 demonstrates this situation on a typical road scene. The figure shows two synthetic frames of a sequence seen from the ALV while it is approaching an intersection. The shaded area in the center represents the region of possible FOE-locations (i.e., the Fuzzy FOE). The building on the right seems to diverge away from the FOE while the image of the truck (which is actually moving in 3-D) remains at a constant image location. If the camera and the truck both continued their current motion, then the two vehicles would collide because

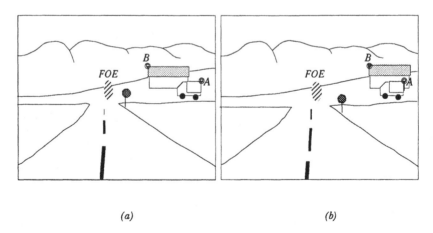

(a)   (b)

Figure 6.7 Indirect detection of a pending collision. Two feature points are tracked in the image in two successive frames *(a–b)*: point A on the (moving) truck, and point B on the building. The stationary part of the image seems to diverge from the FOE region (center), while the point on the truck stays at a constant image location.

point *A* would eventually hit the camera plane.[1]

Applying only the rules for direct motion detection, there is no direct clue to detect the motion of the truck. Instead, the truck would be considered stationary, located at a very large distance in 3-D (because of its zero image motion). However, the motion of the truck could be found by a reasoning chain like the following:

(1) Assume both points *A* and *B* belong to stationary objects.

(2) From their different rates of divergence from the FOE, point *B* must be closer than point *A*.

(3) Since *A* does not show any expansion it would be considered farther away than any other location.

(4) On the other hand, *A* appears to be in front of *B* because *A* is on an object *occluding* the object with feature *B*, which indicates that *A* is *closer* than *B*! This contradicts the conclusion drawn in *(2)*. So either *A* or *B* must be moving, i.e., the original stationary interpretation was wrong.

(5) Since the potential "threat" comes from point *A* moving into the ALV's path, rather than from point *B* moving away, we would conclude that the truck is moving.

The essence is that it is possible to infer 3-D motion by discovering inconsistencies within the stationary parts of interpretations. As an example, the following rule MOTION-FROM-CLOSER responds to a pair of contradicting CLOSER-relationships by asserting the fact (MOVEMENT-BETWEEN...) as a global fact at the root node:

```
❑   (defrule MOTION-FROM-CLOSER
        (CLOSER ?A ?B)
        (CLOSER ?B ?A)
        =>
        (at ROOT (assert (MOVEMENT-BETWEEN ?A ?B))))
```

In reality, there are usually more than just two reference points available,

---

[1] This is, in fact, a well known situation. For example, a rule in marine navigation, mentioned in [68], states that *"If the bearing becomes constant and the distance is decreasing, the two vessels are on collision course."*

such that ambiguities can be resolved more easily. However, it is useful to look for other image clues that can potentially supply information about the 3-D scene layout. One such source could be the analysis of *occlusion*. In the example above, the truck is clearly occluding the building so it must be closer to the camera. Object *recognition* would be another, very powerful source of information. For example, if an object is recognized as a building, any classification of that object – other than a stationary one – can be dismissed. Conversely, an object identified as a truck may potentially be moving, although its current motion may not be apparent.

Lower is Closer

An additional source of information that we have employed is based on a simple heuristic regarding the scene layout likely to be encountered by a land vehicle. Since the vehicle travels upright on some form of surface, the camera looks at a profile in which points are usually closer at the bottom of the image than at the top (Figure 6.8).

The corresponding rule LOWER-IS-CLOSER-HEURISTIC concentrates on image features located in the lower part of the image, particularly features below the FFOE.

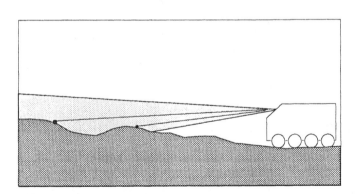

Figure 6.8 The profile of the landscape overlooked by the camera. With few exceptions, features located *lower* in the image generally belong to entities that are *closer* to the camera in 3-D.

❏   **(defrule** LOWER-IS-CLOSER-HEURISTIC
      (CURRENT-TIME ?t)
      (CLOSER ?A ?B)
      (BELOW-FOE ?a ?t)
      (BELOW-FOE ?b ?t)
      (BELOW ?b ?a ?t)
      =>
      (assert (LOWER-CLOSER-CONFLICT-PENDING ?A ?B))).

This is an example of a verification rule that does not fire for every possible feature combination. Instead, only existing CLOSER-relationships are checked if they are compatible with the above heuristic. If there is contradicting evidence, a pending conflict is asserted locally, i.e., within the affected interpretation. Contradicting CLOSER-relationships indicate motion in 3-D. The following pair of rules resolves this conflict by *(a)* asserting that motion has been detected, and *(b)* concluding the *direction* of motion by using an extension of the same heuristic. In each rule, one of the entities involved is hypothesized as being stationary and consequently the other one must be moving. Again the rules are conservative in the sense that only two of the three possible hypotheses are created:

❏   **(defrule** RESOLVE-LOW-CLOSE-CONFLICT-A
      (LOWER-CLOSER-CONFLICT-PENDING ?A ?B)
      (STATIONARY ?A)
      =>
      (assert (MOBILE ?B))
      (assert (MOVES-RECEDING ?B)))

❏   **(defrule** RESOLVE-LOW-CLOSE-CONFLICT-B
      (LOWER-CLOSER-CONFLICT-PENDING ?A ?B)
      (STATIONARY ?B)
      =>
      (assert (MOBILE ?A))
      (assert (MOVES-APPROACHING ?A)))

In summary, as a result of a CLOSER-conflict, the two entities involved are hypothesized as being mobile in two separate interpretations. If one of those two entities can be shown as *not* moving, the situation would become unambiguous. For example, assuming that the camera is moving forward, an object moving *away* from the camera (i.e., moving faster than the camera, *receding*) in 3-D should exhibit image motion *towards* the FOE, or at least no image motion *away* from the FOE. Consequently, if a hypothesis contains an entity A that is receding and its image a is positively found to be *diverging* from the FOE, that hypothesis must be wrong:

❏   (**defrule** REJECT-RECEDING-MOTION
      (MOVES-RECEDING ?A)
      (DIVERGING-FROM-FOE ?a ?t)
      =>
      (**poison**)).

The directive "poison" in the action part of this rule effectively removes the afflicted interpretation. This and other mechanism that alter the structure and contents of the Qualitative Scene Model are covered at a more detailed level in the following chapter.

7

THE QUALITATIVE
SCENE MODEL

One of the key goals of image understanding is to come up with descriptions of the observed scenes that are meaningful for the specific application. Consequently, scene descriptions for different applications may look quite different with respect to their structure and contents. For example, an object recognition system may describe the scene simply as a list of labeled image regions. For a geometric reasoning system, the world may consist of straight line segments, blocks, etc., and the geometrical relations between them.

The *Qualitative Scene Model* (QSM) is the central component of the DRIVE motion understanding engine (see Figure 6.1). The QSM is a 3-D scene description which, unlike many other schemes, describes the behavior of its entities and the relationships between them in very coarse, qualitative terms. The spatial layout of the scene is not stored as a precise geometric description of the scene, but rather its topographic structure relative to the observer is maintained. Since the topography of a stationary scene does not change while it is traversed by the observer, there is no need to continuously update this model as the camera moves forward. Similarly, the 3-D motion of other objects in the scene is also captured in qualitative terms.

One of the most powerful aspects of the QSM is its capability to keep track of multiple scene interpretations, thus reflecting the ambiguities related to 3-D motion analysis. In this chapter, we first describe the primitive elements of the QSM and the representation of multiple scene interpretations. We then look at how conflicts caused by contradicting pieces of evidence are resolved in the QSM and how the model evolves over successive frames.

7.1 BASIC ELEMENTS OF THE MODEL

The QSM is built as a blackboard-like system [52, 53], where a central pool of data represents the current state of the model (Figure 6.2). The entries in this model are associated with various aspects of motion understanding. In particular, the QSM contains the following types of information:

- *Correspondence data* are located at the lowest level of the data hierarchy, generated from the original image data by locating and tracking image tokens from frame to frame. Each image feature is given a unique number when it is observed the first time. For each image feature a of a frame t, a fact is created that includes the image position of the observed feature: (FEATURE $a\ t\ x\ y$).

- *FOE data* are produced by the FOE computation from the given correspondence data (described in Chapter 5), to separate the effects of camera motion into its rotational and translational components. The result is the Fuzzy FOE (i.e., the set of potential FOE-locations), the corresponding rotation angles, and the *derotated* displacement field. Every derotated displacement vector is supplied with additional information about its location relative to the FFOE.

- *Scene entities* are the 3-D counterparts of the 2-D *features* observed in the image. For example, the point feature a located in the image at x, y at time t, denoted by the fact (FEATURE $a\ t\ x\ y$), has its 3-D counterpart in the model as (MEMBER A). Model entities carry the same identification number as their corresponding image features (e.g., $a = A = 145$). Scene entities are the basic building blocks for creating scene interpretations. Hypotheses about the 3-D properties of entities are expressed in the form of facts like (STATIONARY A), (MOBILE B), (CLOSER $X\ Y$), etc.

- *Control data* are used to enforce certain actions in the course of updating the QSM. Some of these entries control the selective activation and deactivation of groups of rules, while others are necessary for conflict resolution. A typical example is the assertion (LOWER-CLOSER-CONFLICT-PENDING ...) which indicates that a particular conflict is pending that should be resolved by firing appropriate rules (as mentioned in Chapter 6).

7.2 REPRESENTING MULTIPLE INTERPRETATIONS

Multiple Worlds

Due to the loss of 3-D spatial information by the imaging process, the analysis of dynamic (and static) scenes is inherently ambiguous. This means that more than one interpretation of given image sequence may be feasible and even plausible at the same time. Consider the following example:

Two image points A and B have been observed and are considered to be *stationary* up to some point in time t_i. At this point, the QSM would contain the facts

(MEMBER A), (MEMBER B), (STATIONARY A), (STATIONARY B).

At t_i, it is concluded from some image events that the entities A and B must be moving relative to each other in 3-D. Consequently, at least one of the two entities must be moving. We furthermore assume that it cannot be determined from the available evidence which of the two entities (if any) is stationary. The solution is to hypothesize two separate interpretations, each taking care of one of the possible cases. We can view these interpretations as different possible *worlds* W_1–W_3 that contain the following assertions:

W_1: (MEMBER A) W_2: (MEMBER A) W_3: (MEMBER A)
      (MEMBER B)   (MEMBER B)   (MEMBER B)
      (MOBILE A)   (STATIONARY A)   (MOBILE A)
      (STATIONARY B)   (MOBILE B)   (MOBILE B)

All of the three worlds are feasible and, in the absence of any additional information, also plausible. W_3, where both A and B are mobile, could be considered less likely than the other interpretations. In fact, the actual implementation would not have created W_3 at all in this particular situation.

When rules access the data contained in the model, every fact must be viewed in the *context* of the world in which it is valid. In the previous example, the fact (MOBILE A) is only true in worlds W_1 and W_3. Similarly, the fact (STATIONARY B) only holds in W_1. If a rule fires on a pattern valid in a particular world, then the validity of its conclusions is also limited to the context of that world. Suppose there is a rule BOGUS that is defined by

```
(defrule BOGUS
   (STATIONARY ?X)
   (MOBILE ?Y)
   =>
   (assert (FOO ?X ?Y))).
```

Applied to the above knowledge base consisting of W_1-W_3, the rule would fire two times, once in W_1 and once in W_2, while world W_3 does not contain the matching facts and remains unaffected. Notice that the new facts are asserted within the interpretation that contains the matching premises. The result is:

W_1:	(MEMBER A)	W_2:	(MEMBER A)	W_3:	(MEMBER A)
	(MEMBER B)		(MEMBER B)		(MEMBER B)
	(MOBILE A)		(STATIONARY A)		(MOBILE A)
	(STATIONARY B)		(MOBILE B)		(MOBILE B)
	(**FOO** $B\,A$)		(**FOO** $A\,B$)		

Of course, as more entities are contained in the QSM, the number of possible scene interpretations grows combinatorially. Explicitly storing each individual scene interpretation at any point in time would result in exponential space requirements. Fortunately, it is possible to avoid this problem to a great extent by structuring the set of interpretations such that the facts shared by multiple interpretations are stored together in a single place.

The Interpretation Graph

Reasoning with multiple worlds is a well known strategy for truth maintenance in Artificial Intelligence. The QSM is structured as a directed graph whose nodes contain *partial scene interpretations*. Each partial interpretation stands for a hypothesis represented by a consistent collection of assertions. Every node of this *interpretation graph* (IG), except the single root node, inherits the facts valid in its superior node(s).

Viewpoints

The *viewpoint* mechanism in *ART* is a convenient abstraction of hypothetical reasoning in a rule-based environment. It allows the manipulation of parallel alternatives without the usual bookkeeping required from the programmer. Alternatives are arranged as a single directed graph structure, in which every node represents a different *viewpoint*. Every node – except the root node – inherits the facts valid in its superior node(s). The root node itself holds all those facts that are universally true in any possible context.

The construction of the QSM starts at the root node by asserting one fact (MEMBER *i*) for every entity *i* observed in the image. The second step is to create a *default* interpretation for the set of new entities in the QSM. Unless otherwise known, the default assumption is that every entity is part of the stationary environment. This is accomplished by creating an individual viewpoint for each entity contained in the root, by using the following rule:

❑   (**defrule** DEFAULT-INTERPRETATION
     (MEMBER ?x)
     =>
     (**hypothesize** (**assert** (STATIONARY ?x))))

This rule fires at the root once for every member of the model. Figure 7.1 shows a simple IG for a scene with 4 entities that are all believed to be stationary (by default). A look *inside* one of these new viewpoints (e.g., node I-3) would reveal that the following facts are (locally) true:

    (MEMBER 1), (MEMBER 2), (MEMBER 3), (MEMBER 4),
    (STATIONARY 2).

Any rule whose antecedents are a subset of these facts could fire within node I-3. Notice that, at this point, there exists no *complete interpretation*, i.e., a single node that contains (or inherits) a classification for every entity in the scene. Fortunately, updating the QSM can be accomplished locally on partial interpretations and does not require complete interpretations. Later we show how complete interpretations are created by merging partial interpretations.

Augmenting Partial Interpretations

Assume, for example, that the rule DIRECT-PAIR-MOTION (see Chapter 6) has determined that the two entities 1 and 2 must be moving relative to each other in 3-D, but could not determine which one was moving. The firing of the rule would cause the new fact (MOVEMENT-BETWEEN 1 2) to be asserted at the root node I-1 of the IG (Figure 7.2). The model must now be updated to at least eliminate any interpretation that considers *both* entities 1

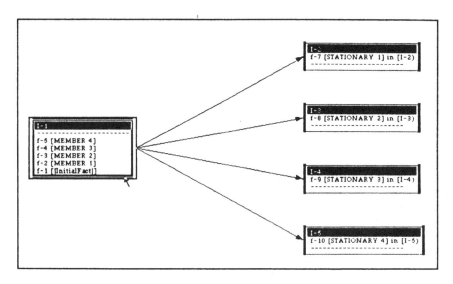

Figure 7.1 Simple interpretation graph. The model contains four entities (labeled 1...4), which are initially assumed to be stationary (by default). Each entity *i* is recorded as a member of the current model in the root node I-1 by (MEMBER *i*). For each entity, a *partial interpretation* has been created (represented by child nodes I-2 to I-5) in which only (STATIONARY *i*) is true.

and 2 stationary, as accomplished by the following pair of rules:

❏ (**defrule** RELATIVE-MOTION-A
     (MOVEMENT-BETWEEN ?A ?B)
     (STATIONARY ?A)
     =>
     (assert (MOBILE ?B)))

❏ (**defrule** RELATIVE-MOTION-B
     (MOVEMENT-BETWEEN ?A ?B)
     (STATIONARY ?B)
     =>
     (assert (MOBILE ?A)))

A verbal interpretation of the first rule would be: *"If a 3-D movement has been observed between entities A and B then, assuming that A is stationary, entity B should be mobile."*

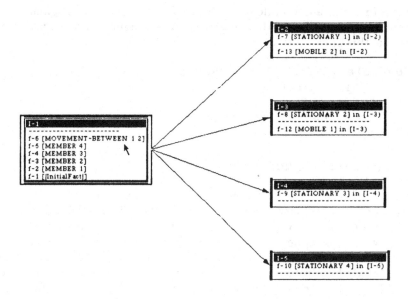

Figure 7.2 Local updating of partial hypotheses. The fact (MOVEMENT-BETWEEN 1 2) has been asserted at the root node I-1. As a consequence, rule RELATIVE-MOTION-A fires in node I-2 and RELATIVE-MOTION-B in node I-3, leaving behind the new facts (MOBILE 2) in I-2 and (MOBILE 1) in I-3. Notice that within nodes, original hypotheses are shown above a dashed line and the conclusions are shown below that line.

When trying to find a matching pattern of facts to fire these rules, the rule interpreter in *ART* searches for the least specific partial interpretation (i.e., the node closest to the root), where all the antecedents are satisfied. By definition, the rule will then put the new assertions into this particular node. In Figure 7.2, the rule RELATIVE-MOTION-A fires in node I-2, asserting the new fact (MOBILE 2) at that node. Notice that the fact (MOVEMENT-BETWEEN 1 2) is inherited from the root node. Similarly RELATIVE-MOTION-B fires in node I-3, asserting (MOBILE 1) there.

Merging Partial Interpretations

Partial interpretations (nodes) may be merged automatically by the inference engine whenever a rule requires a *conjunction* of assertions located in separate nodes. In this case, *ART*'s viewpoint mechanism automatically attempts to create a new node that inherits the facts from the original nodes. Assume the state of the QSM is the one shown in Figure 7.1 and that we had a rule like this one:

❏ (**defrule** VERY-SPECIAL-MERGE
        (STATIONARY 1)
        (STATIONARY 2)
        (STATIONARY 3)
        (STATIONARY 4)
     =>
        (**assert** (COMPLETE)))

The aim of this rule is to find a partial interpretation where all entities 1...4 are considered stationary. In the current model (Figure 7.1), there is no such node that satisfies all the preconditions. What the viewpoint mechanism now does, is trying to combine existing hypotheses in order to create such a node. Figure 7.3 shows the result of this attempt. In order to build the final node I-8, the viewpoint mechanism creates two intermediate nodes I-6 and I-7. After the nodes have been merged, the rule eventually fires in node I-8 and asserts the new fact (COMPLETE).

While the automatic merging feature saves a lot of bookkeeping on the programmer's side, it has the potential for producing inconsistent interpretations. For example, if we apply the rule VERY-SPECIAL-MERGE at the state shown in Figure 7.2, the viewpoint mechanism would first merge nodes I-2 and I-3. This would lead to the situation shown in Figure 7.4. The newly created node I-6 contains the facts

(STATIONARY 1), (STATIONARY 2), (MOBILE 1), (MOBILE 2).

Since an entity cannot be considered stationary *and* mobile within the same interpretation, node I-6 is inconsistent.

Poisoning Partial Interpretations

The problem in Figure 7.4 is handled by a set of local conflict resolution rules, which detect inconsistent nodes and remove (*poison*) them permanently, e.g.:

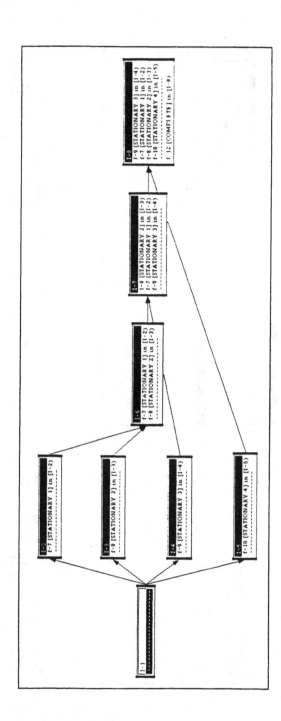

Figure 7.3 Automatic merging of viewpoints. Partial interpretations are merged automatically in order to satisfy the antecedents of a rule. With the QSM in the state shown in Figure 7.1, the attempt to fire rule VERY-SPECIAL-MERGE causes the successive combination of existing nodes. The rule eventually fires at the final node I-8 where it asserts the new fact (COMPLETE).

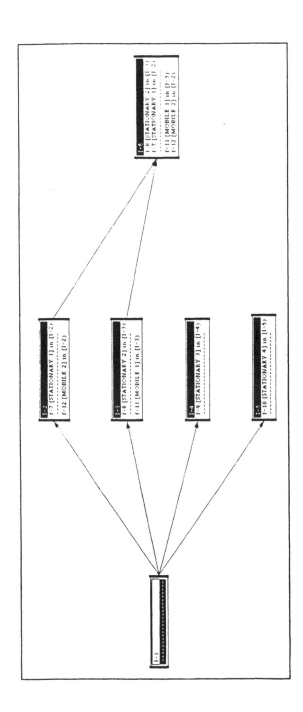

Figure 7.4 Inconsistent hypotheses created by automatic merging of viewpoints. Automatic merging of viewpoints may lead to inconsistent hypotheses. With the QSM in the state shown in Figure 7.2, the attempt to fire rule VERY-SPECIAL-MERGE causes the creation of a new node I-6 which contains the conflicting facts (STATIONARY 1) & (MOBILE 1) and (STATIONARY 2) & (MOBILE 2).

❑ **(defrule** REMOVE-STATIONARY-AND-MOBILE
    (STATIONARY ?A)
    (MOBILE ?A)
  =>
    (poison))

The execution of this rule will cause the affected node and all its subordinate nodes to be removed from the interpretation tree. The viewpoint mechanism takes care of this by *never* recreating a node that was poisoned once. While this kind of inconsistency of the model can be resolved by purely local actions, more complex decisions are necessary for other types of conflicts.

Universal Truth

There are cases when a predicate of an entity or a relationship between entities can be stated with certainty, irrespective of the particular context. For example, an image point that is moving *towards* the FOE can only be caused by an entity moving in 3-D (see Section 6.4), regardless of the properties of other entities.

In order to communicate this information to all current interpretations in the model, the fact (MOBILE *A*) is placed at the *root* of the viewpoint structure. Since there is only one root node, any viewpoint inherits all facts asserted at the root by default. This may, of course, lead to inconsistencies within the viewpoints, which have to be resolved in a suitable manner. This may have the desirable effect of eliminating entire branches of the interpretation graph, thus reducing the number of possible scene interpretations.

Assembling Complete Interpretations

Since the QSM is organized as a network of partial interpretations, there is normally no explicit listing of the possible interpretations of the entire scene, called *complete* scene interpretations. A node of the interpretation graph is said to be a *complete* interpretation, if it can supply a STATIONARY or MOBILE label, either locally or by inheritance, for each entity in the scene.

Most of the time, the development of the QSM proceeds by adding, modifying, and removing partial hypotheses without the need to explicitly assemble complete interpretations. Queries like *"Could entity X be mobile in any current scene interpretation?"* can be answered by evaluating only *partial* interpretations.

Whenever needed, *complete* interpretations are assembled on demand by combining all possible (and permissible) combinations of the partial interpretations. The following simple rule initiates the necessary merges and marks the resulting nodes with the fact (COMPLETE):

❑ **(defrule** ASSEMBLE-COMPLETE-INTERPRETATIONS
    (forall (MEMBER ?A)
       (STATIONARY | MOBILE ?A))
  =>
    (assert (COMPLETE))).

Figure 7.5 shows the result of applying this rule to the interpretation graph of Figure 7.2. Nodes I-8 and I-10 have been marked as complete interpretations; I-7 and I-9 are intermediate nodes created by the automatic merging process. Inconsistent intermediate nodes that may have been created by this process were removed by local resolution rules.

Notice that poisoning a viewpoint not only removes the afflicted viewpoint node itself, but also all its subordinate viewpoints. Therefore, any *complete* interpretation which inherited information from the afflicted viewpoint is removed from the model as well. This is perfectly consistent with the concept of partial interpretations. Whenever a detail of an interpretation can be proven wrong, the complete interpretation is not feasible either.

Evaluating Scene Interpretations

The information contained in the QSM is supposed to aid the ALV to perform its mission. While the number of possible scene interpretations is usually small, it may sometimes be necessary to select and accept a *single* scene interpretation as the basis for further actions. In the simplest case, only one scene interpretation exists. When several scene interpretations are feasible at the same time, it is necessary to evaluate and rank the individual interpretations. Some of the possible ranking criteria are, among others:

•   the number of entities considered *stationary* in an interpretation;

•   the time elapsed since the interpretation was created;

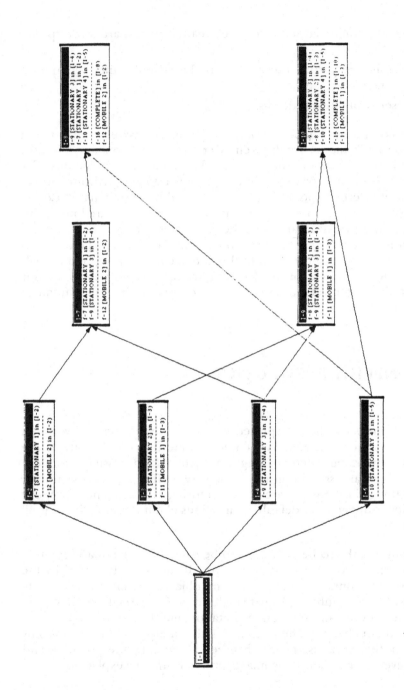

Figure 7.5 Assembling *complete* scene interpretations. Two complete interpretations, I-8 and I-10, have been created from the situation shown in Figure 7.2. Notice that the nodes I-2 and I-3 cannot be merged because the partial interpretations represented by those viewpoints are incompatible.

- the average time elapsed since the creation of its partial interpretations;
- the number of partial interpretations shared with other (complete) interpretations;
- the absence of unlikely details.

The actual criteria used depend upon the particular task environment. It may even be desirable to use different scene interpretations for different kinds of action or extract information shared by a set of interpretations. For instance, the estimation of vehicle motion relative to the environment requires a set of reference point that can be trusted to be stationary. Consequently, we may want to know those points that are *not* considered mobile in *any* existing interpretation and have been observed for a certain period. This would not even require the creation of complete interpretations. On the other hand, a task concerned with threat detection may respond to each entity considered mobile in any current scene interpretation. In our implementation, interpretations were ranked only according to the number of stationary entities.

7.3 CONFLICT RESOLUTION

Motion sequences can be interpreted in many different ways. Even in a completely static environment, we could assume that every entity in the scene is moving independently along a particular (and possibly complex) trajectory. Such an assumption would be very inefficient, because usually only a few things in the scene are actually moving around. Therefore, our system relies strongly upon default assumptions about the mobility of individual entities.

At the beginning, the scene is thought to be completely stationary, creating a single interpretation of the environment. If the scene traversed by the camera does not contain any other moving objects, any image motion observed must be compatible with that single scene interpretation. New entities are added continually to that interpretation, but the number of interpretations would not change. The structure of the interpretation graph is only altered when there is a discrepancy between an existing interpretation and the image events, i.e., when some image motion cannot be explained.

Local Conflict Resolution

The content of an interpretation can be seen as a set of *hypotheses* and a (possibly empty) set of *conclusions* that follow from the hypotheses:

$$\{hypothesis\text{-}1, hypothesis\text{-}2,... \} \Rightarrow \{conclusion\text{-}1, conclusion\text{-}2,... \},$$

e.g., $\{(\text{STATIONARY } 1)\} \Rightarrow \{(\text{MOBILE } 2)\}$ in node I-2 of Figure 7.2. A conflict occurs in an interpretation, when the expression formed by the conjunction of hypotheses and conclusions[1] is false, i.e.

$$\neg \{hypothesis\text{-}1 \wedge hypothesis\text{-}2 \wedge ... conclusion\text{-}1 \wedge conclusion\text{-}2...\}.$$

This is the case in Figure 7.4, where node I-6 was created by merging nodes I-2 and I-3 :

$$\neg \{(\text{STATIONARY } 1) \wedge (\text{STATIONARY } 2) \wedge (\text{MOBILE } 1) \wedge (\text{MOBILE } 2)\}.$$

Assuming that the inferences from premises to conclusions are correct, a conflict indicates that the hypotheses of the affected interpretation must be invalid. Therefore, it is a legitimate move to eliminate this interpretation from the model by poisoning it. In case of the *stationary/mobile*-conflict, no other action is required because the integrity of the whole model is not affected.

In Section 7.2 we showed how inconsistent partial interpretations are removed from the QSM by executing the constraint rule REMOVE-STATIONARY-AND-MOBILE, which simply *poisons* that particular node and all its successor nodes. This has been referred to as *local* conflict resolution, because the action is executed within a particular interpretation, regardless of the global state of the model.

Global Conflict Resolution

When conflicts can be resolved locally, the modifications to the QSM are confined to that partial interpretation where the conflict occurred. In some

---

[1]   In *ART*, hypotheses and conclusions within a viewpoint node are not distinguished logically, i.e., they are all equally visible to the rules as facts. For example, there is no way to write a rule that responds only to a hypothesis.

cases, however, the detection of incorrect premises may lead to consequences beyond the afflicted viewpoint. In particular, when premises that are *inherited* from superior viewpoints are proven false, actions must take place where the false facts were asserted first. Consequently, such an action can potentially change the entire structure of the interpretation graph.

One example for global effects is the direct detection of 3-D motion. When an entity A is known to be moving in 3-D, the system would assert the fact (MOBILE A) at the root node of the interpretation graph. By doing so, the new fact becomes visible in all other partial interpretations (i.e., viewpoint nodes) in the graph. This again may lead to conflicts inside the partial interpretations which are subsequently resolved by local operations.

Meta Rules

Conflicts are essential for the operation of the QSM. Any *feasible* interpretation, i.e., one that cannot be proven false, may also be correct. Consequently, the number of interpretations grows exponentially with the number of scene entities, unless we have heuristics that limit the number of alternatives that are pursued in parallel. These heuristic guidelines are summarized as the following meta rules:

• Always tend towards the "most stationary" (i.e., most conservative) solution. The default assumption of all entities being stationary reflects this principle.

• Assume that an interpretation is feasible unless it can be proven to be false (i.e., a strategy of *least commitment*).

• Findings from observations in the image are interpreted in the context of each individual viewpoint. The actual conclusion depends on the contents of this viewpoint.

• If a new fact causes a conflict in one interpretation, but the same fact can be accommodated without conflict by another current interpretation, then only remove the conflicting interpretation.

• If a new fact cannot be accommodated by *any* current interpretation, create a new, plausible interpretation which is conflict-free and remove the conflicting ones (prune the search tree).

7.4 DYNAMIC EVOLUTION OF THE QSM

At any point in time, the state of the QSM is defined by the structure of the interpretation graph and the contents of its nodes. In the following, we use an example to illustrate the dynamic behavior of the QSM over a short period of time.

Tracing Complete Interpretations

The scene model shown in Figure 7.6 contains three entities A, B, C. Initially, nothing is known about the relationships between those three entities and whether they are stationary or mobile. By default, they are assumed to be stationary. The initial interpretation of the scene thus contains only a single interpretation W_1:

$W_1(t_0)$: (STATIONARY A)
(STATIONARY B)
(STATIONARY C)

The flow graph in Figure 7.6 symbolically shows the states of the model as it develops over time. Entities considered stationary are drawn as squares, mobile entities are drawn as circles.

Suppose that between t_0 and t_1 all three points show some divergence away from the FOE, such that a valid conclusion would be that A is *closer* to the camera than B, A is *closer* than C, and C is *closer* than B. From the information gathered up to this point, the complete interpretation W_1 at time t_1 looks like this:

$W_1(t_1)$: (STATIONARY A)
(STATIONARY B)
(STATIONARY C)
(CLOSER A B)
(CLOSER A C)
(CLOSER C B)

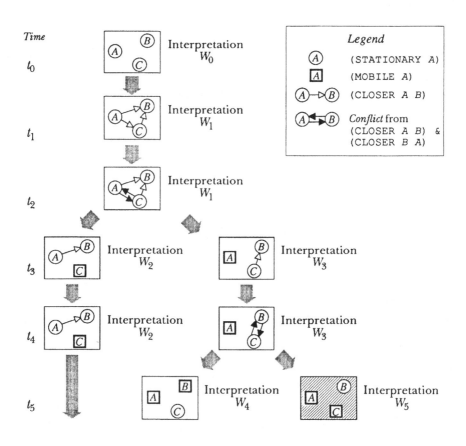

Figure 7.6 Evolution of complete interpretations in the QSM. Scene interpretations are shown as boxes. At time t_0 three entities A, B, C exist in the model, which are initially assumed to be stationary in the single interpretation W_0. At time t_1, three CLOSER-pairs are found among A, B and C, indicated by arcs in interpretation W_1. At time t_2, a conflict occurs in W_1 through the contradictory facts (CLOSER A C) and (CLOSER C A). Two new interpretations (W_2 and W_3) are created, each containing one feature considered mobile (C and A, respectively). At time t_4, a new conflict occurs in interpretation W_3 from the additional fact (CLOSER B C). As before, interpretation W_3 is forked into two new interpretations W_4 and W_5. However, the branch W_5 is pruned because at the same time another interpretation exists (W_2) that already regards C as mobile. Therefore, only interpretations W_2 and W_4 are maintained beyond t_5.

Now suppose that at time t_2, one of the rules claims that C is closer than A and tries to assert this fact into the current interpretation. Clearly, the new interpretation would contain the contradicting facts (CLOSER A C) and (CLOSER C A), which would therefore not be a feasible interpretation. The conflict is resolved in the usual way by creating two disjoint interpretations W_2 and W_3, which consider either A or C as mobile:

$W_2(t_3)$:	(STATIONARY A)	$W_3(t_3)$:	(MOBILE A)
	(STATIONARY B)		(STATIONARY B)
	(MOBILE C)		(STATIONARY C)
	(CLOSER A B)		(CLOSER C B)

At this point in time (t_3), *two* interpretations of the scene are feasible simultaneously. All active interpretations are pursued until they enter a conflicting state, in which case they are either branched into new interpretations or removed from the QSM. Since the CLOSER relationship is only meaningful between stationary entities, hypothesizing an entity as being mobile causes the removal of all CLOSER-relationships that refer to this entity.

At time t_4, a rule claims that if both B and C were stationary, then B must be *closer* than C. This would not create a conflict in interpretation W_2, where C is considered mobile anyway and would make (CLOSER B C) meaningless.

The second interpretation W_3 cannot ignore the new conclusion (CLOSER B C) because it currently considers both entities B and C as stationary. However, it also contains the contradicting fact (CLOSER C B)! Again, interpretation W_3 would be branched into two new interpretations, with either (MOBILE B) or (MOBILE C), respectively. At this point, we find three feasible scene interpretations, W_2, W_4, and W_5.

Viewpoint Dynamics

With the approach just described, the set of possible scene interpretations is generally subject to combinatorial growth. In a real situation with several dozens of entities or more, the computational requirements for handling such a model would become impractical. Using the same example as before, we now illustrate how the structuring of interpretations into partial interpretations and the application heuristics can avoid this sort of combinatorial explosion (Figures 7.7–7.11). Figure 7.7 shows the state of the QSM at time $t = t_1$, with three entities and three CLOSER-pairs (see Figure 7.6).

The first conflict at time t_2 is caused by the assertion of (CLOSER A C), shown in node I-5 of Figure 7.8a. The conflict suggests that at least one fact of the conditions that interpretation I-5 is based on, namely (STATIONARY A) and (STATIONARY C), must be false. Simply removing the conflicting interpretation I-5 would not automatically create this conclusion. The problem is solved indirectly by the rule MOTION-FROM-CLOSER (see Section 6.4) which detects the conflict in I-5 and asserts the fact (MOVEMENT-BETWEEN A C)[1] at the root node I-1 (Figure 7.8b). In reaction to the new facts contained in the root node, the rules RELATIVE-MOTION-A and RELATIVE-MOTION-B (Section 7.2) fire and assert the facts (MOBILE C) in I-2 and (MOBILE A) in I-4 (Figure 7.8c).

As a consequence of the resulting CLOSER-conflict, interpretation I-5 is eliminated by rule REMOVE-STATIONARY-AND-MOBILE. Now the original conflict that occurred at time t_2 is finally resolved (Figure 7.9). The nodes I-6 and I-7 contain two separate scene interpretations corresponding to W_2 and W_3 at time t_3 in Figure 7.6.

The second conflict in this example is caused by the assertion of the fact (CLOSER B C) in node I-6 of Figure 7.10 (at time t_4). As in the previous case, two new facts (MOVEMENT-BETWEEN B C) and (MOVEMENT-BETWEEN C B) are asserted in the root node I-1, which immediately lead to the elimination of node I-6 (Figure 7.11). At the same time, (MOBILE B) is asserted at node I-4. Because I-4 explains every entity of the model and is consistent, it represents a feasible and complete scene interpretation. The same holds for node I-7.

Comparing the situation depicted in Figure 7.11 and the state of the model at time t_5 in Figure 7.6, we can observe the following: Node I-7 corresponds to W_2 in Figure 7.6 and node I-4 corresponds to W_4. Notice that no interpretation was created that corresponds to W_5. This is due to the implicit heuristic that new hypotheses are created by trying to make the smallest possible modifications to existing ones. In this case, there was no need to create interpretation W_5 because C is already *mobile* in W_2 (node I-7). Thus, only the interpretations corresponding to nodes I-4 and I-7 are maintained beyond t_5.

---

[1]   Because of the symmetry in the rule's left-hand side, the rule MOTION-FROM-CLOSER actually fires twice in node I-5 and also asserts the mirrored fact (MOVEMENT-BETWEEN C A) at the root node.

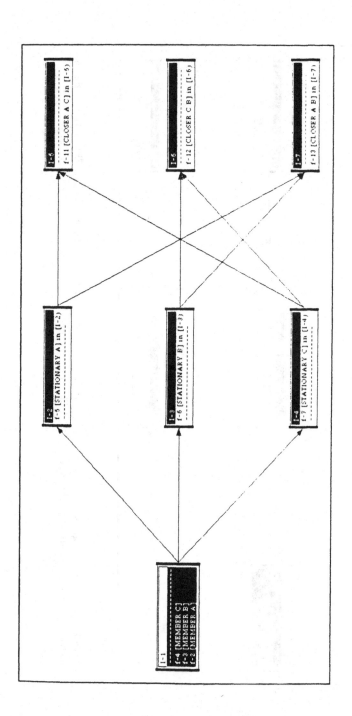

Figure 7.7 Interpretation graph at time t_1. All three entities A, B, C are initially considered stationary, contained in the viewpoint nodes I-2, I-3, and I-4. Closer-relationships are represented by nodes I-5, I-6, and I-7. Only one complete interpretation is possible (which is not assembled at this point).

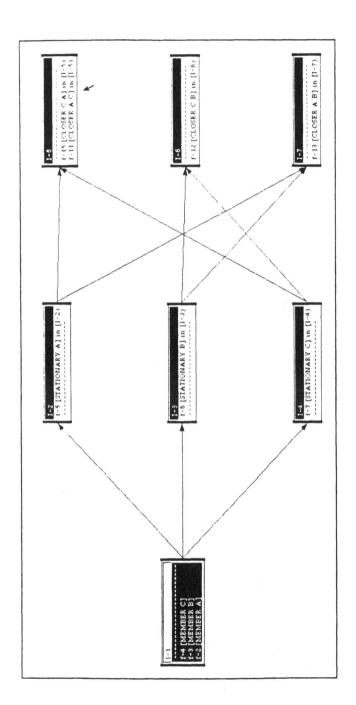

Figure 7.8 Interpretation graph at time t_2. (*a*) A conflict arises due to the assertion of the fact (CLOSER C A) in node I-5, which contradicts the existing hypothesis (CLOSER A C).

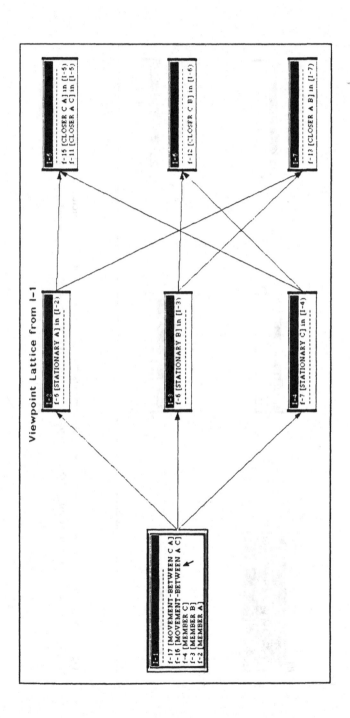

Figure 7.8 (*continued*) (*b*) The conflict in I-5 has been detected by the rule MOTION-FROM-CLOSER, which asserts its conclusions (MOVEMENT-BETWEEN *A C*) and (MOVEMENT-BETWEEN *CA*) at the root node I-1.

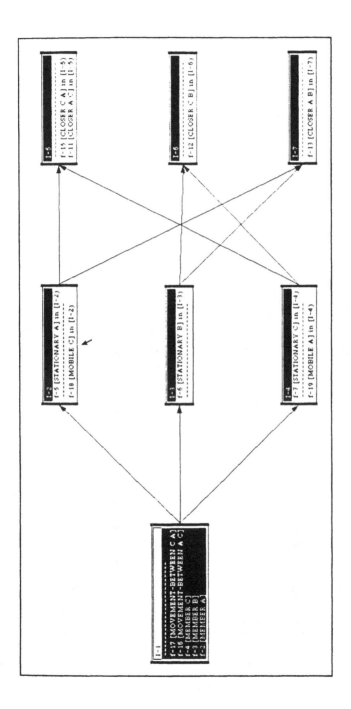

Figure 7.8 (*continued*) (*c*) As a consequence of the previous conflict, entities *A* and *C* are concluded to be *mobile* in nodes I-4 and I-2 (*arrow*). The pending conflict is not completely resolved yet,because node I-5 is still inconsistent.

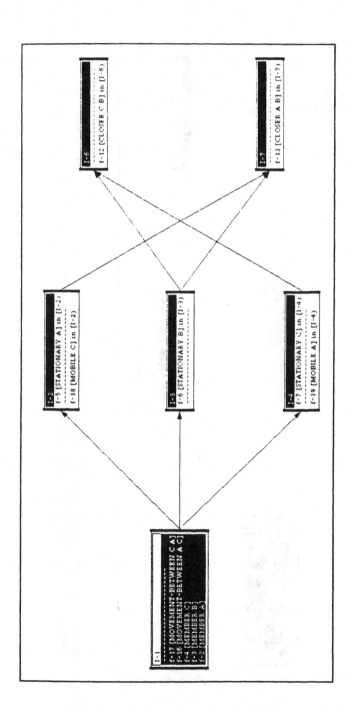

Figure 7.9 Interpretation graph at time t_4. The afflicted node I-5 has been removed and two complete scene interpretations remain: I-6 and I-7, which correspond to interpretations W_3 and W_2, respectively, in Figure 7.6.

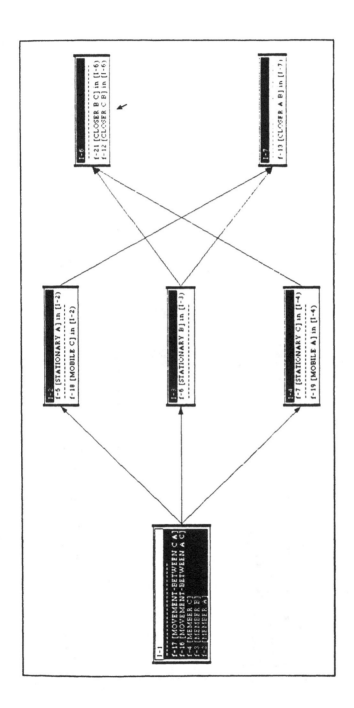

Figure 7.10 Interpretation graph at time t_4. A new conflict arises due to the assertion of the fact (CLOSER B C) in node 1-6, causing again (MOVEMENT-BETWEEN B C) and (MOVEMENT-BETWEEN C B) to be asserted at the root node.

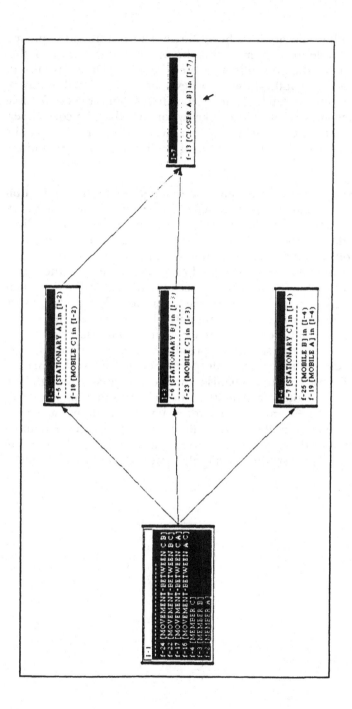

Figure 7.11 Interpretation graph at time t_5. The inconsistent node I-6, has been eliminated. (MOBILE B) and (MOBILE C) are concluded in I-4 and I-3. Nodes I-4 and I-7 represent the two complete scene interpretations W_4 and W_2, respectively, of Figure 7.6.

Obviously, the result would not have been the same for a different sequence of image events. While this scheme limits the number of current hypotheses, there is, of course, the possibility that the true interpretation is missed. Given a set of feasible hypothesis, we can never be certain whether any of them is the correct interpretation or not. Of course, when the set of feasible interpretations becomes empty, we also know for sure that the correct version has been missed. In this case, the system must perform a cleanup of its partial interpretations and create a new set of feasible scene interpretations to continue.

In summary, the central aspects of our scene model are the representation of 3-D properties in *qualitative* terms, as well as its capability to maintain multiple scene interpretations simultaneously. Of course, the model is not limited to qualitative descriptions. For example, we could keep relative distance measurements from stationary scene entities not only as a partial ordering in depth but could also include the estimated ranges of values into the representation. Similarly, the motion of moving scene entities could be described in much more precise terms as we have done so far, possibly including hypotheses about acceleration and the complexity of their trajectories in 3-D space. The model can thus be refined in accordance to the accuracy of the available input data. All this should be possible within the original framework of representing multiple scene interpretations by a network of partial interpretations. The viewpoint mechanism that we have used, although convenient for building a prototype, is neither the only way to implement this concept nor the most efficient one. However, implementation issues have not been a major concern in this work. In the following chapter, we demonstrate the operation of the qualitative reasoning mechanism and the evolution of the QSM on both a synthetic example as well as on a real ALV image sequence.

8

EXAMPLES

The preceding chapters covered most of the details relevant to our qualitative motion understanding approach, in particular the *Fuzzy Focus of Expansion* and the *Qualitative Scene Model*. In the following, we show the operation of this concept with two major examples using image sequences that are typical for autonomous land vehicle navigation. The first experiment was performed with an animated image sequence in order to have the ground truth and the vehicle motion parameters available. This setup was used to develop and debug the reasoning mechanism. The second example shows the results from a real image sequence taken from the moving ALV.

8.1 SIMULATED DATA

For this experiment, a sequence of images was generated which is similar to the ones typically encountered by the ALV. The images were computed from a simple 3-D model of the scene which contains several stationary and moving objects. The focal length of the camera and its orientation relative to the vehicle correspond to the real setup. The ALV itself is moved through this environment by changing its position and orientation in world coordinates after each frame. Thus, the translation and rotation parameters of the ALV's own motion were known at any time.

This approach has several advantages:

- Since the actual 3-D layout of the scene is known and the motion of the ALV and moving objects is under control, the results can be verified directly.

- The efficiency of various knowledge sources (rules) can be evaluated by simply altering the layout of the scene.

- Different driving modes and independent object motion can be simulated for which real data are not available.

Figure 8.1 shows the image sequence used for these experiments. During this sequence, the ALV is moving towards an intersection which is located at a distance of roughly 80 meters in the first frame ($t = 0.0$). The horizontal viewing angle is about 40°, the depression angle of the camera is 17°. During each frame period of 0.5 seconds, the ALV moves forward by a distance of 2 meters, corresponding to a speed of about 15 $^{km}/h$. The vehicle also performs random horizontal and vertical rotations in the range of ±3°.

The scene contains two independently moving objects. One other vehicle (marked **F**) is approaching the ALV on the same road at a constant speed of 22 $^{km}/h$. A second car (marked **P**) is crossing the path of the ALV from right to left at a constant speed of 36 $^{km}/h$. It enters the field of view in the second frame ($t = 0.5$), when the distance to the ALV is about 80 meters. The initial distance of the camera from the pole marked **M** is 17 meters; the hills in the background are 600 meters away.

The last picture of Figure 8.1 (in the lower right-hand corner) shows the traces of the points labeled **A** to **P** from all frames. This should only visualize the effects of the ALV's rotation – the FOE-algorithm was not applied to this sequence. The main purpose was to evaluate the construction of scene interpretations within the *Qualitative Scene Model*.

The development of the QSM for this image sequence is shown graphically in Figure 8.2. At four points in time (0.5, 1.0, 2.0 and 4.5 seconds), the state of the QSM is displayed as the set of complete interpretations that existed at these moments. Concurrent interpretations are stacked vertically in

Figure 8.2, but no ranking was applied. Each interpretation labels entities as either *stationary* or *mobile*. Stationary entities are marked with a circle, while all other (non-circular) symbols denote mobile entities.

Inferred CLOSER-relationships between stationary entities are indicated by a connecting line between two image features, the *closer* of the two entities being marked with the larger circle in Figure 8.2. Since the CLOSER-relationship is transitive, the sizes of the circles are adjusted accordingly.

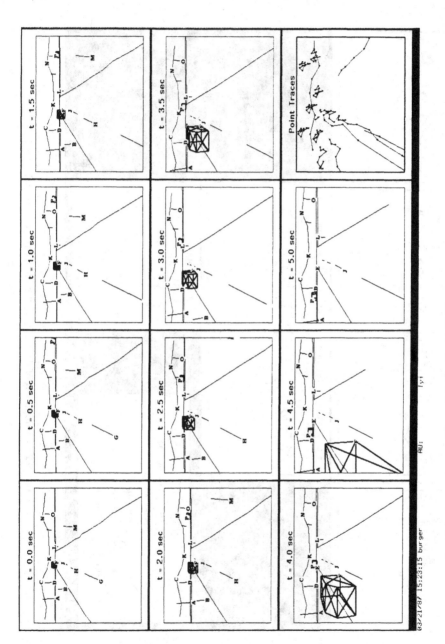

Figure 8.1 Simulated ALV image sequence.

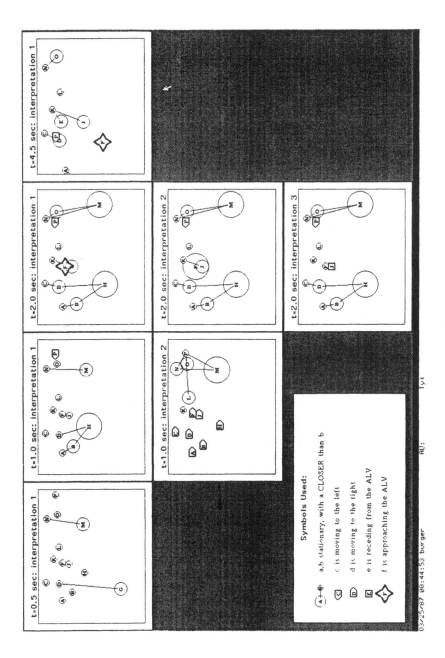

Figure 8.2 Development of the QSM for the sequence in Figure 8.1.

In this example, only *passes* between features were used to hypothesize CLOSER-relationships, not the relative rate of expansion away from the FOE. This explains why no such relationships were found *across* the center of the image.

Instant 1 (t = 0.5)

At time $t = 0.5$ (after one pair of frames), the scene is considered completely stationary, which was the default assumption at the beginning. Two CLOSER-relationships, (CLOSER **G D**) and (CLOSER **M N**), have been established.

Instant 2 (t = 1.0)

At time $t = 1.0$, the moving car (**P**) has entered the field of view. Relative movement across the FOE-area leads to the creation of two scene interpretations. *Interpretation-1* considers **P** as being *mobile* and moving to the left. *Interpretation-2* sees **P** as *stationary* and has concluded CLOSER-relationships with respect to other stationary entities as a consequence of the observed relative movements.

The mobile objects in *Interpretation-2* are seen on the left side of the FOE and they are assumed to be moving to the right. Although *Interpretation-1* is clearly the correct one and would be ranked higher because more entities are stationary, the second interpretation is also feasible at this point in time. *Interpretation-2* is eliminated after the subsequent frame, when one of the features on the left (**B**) is found to be definitely *diverging* from the FOE to the left. This, of course, contradicts the hypotheses of *Interpretation-2* at $t = 1.0$ that **B** is moving to the right.

Instant 3 (t = 2.0)

At time $t = 2.0$, **P** has been definitely identified as being mobile and moving to the left. The approaching van (**F**) has not been detected up to this moment. Since it is approaching the ALV approximately on a straight path with constant speed, its motion is not found directly (see Figure 6.7 for an illustration of this situation). However, feature **F** has been found to be *passing* feature **J**, which leads to the conclusion that if both **F** and **J** were stationary, then **F** must be *closer* to the ALV than **J**. This conclusion is reflected by *Interpretation-2* at $t = 2.0$. This interpretation is *feasible* and the approaching of point **F** could not be detected by strictly geometric reasoning. However, point **J** is *lower* in the image than **F**, and should therefore be *closer* than **F**. This conflict is discovered by one of the heuristic rules (Figure 6.8) which creates two new interpretations. *Interpretation-1* sees **F** approaching and **J** stationary, while *Interpretation-3* sees **F** *stationary* and **J** *receding* from the ALV. In *Interpretation-2*, **F** is still considered stationary but "deep" (i.e., far away) in the scene. Consequently, three scene interpretations are active at $t = 2.0$.

Instant 4 ($t = 4.5$)

Interpretation-3 is rejected when **J** is found to be diverging from the FOE, which contradicts the hypotheses that it is *receding*. Eventually, at $t = 4.5$, only one interpretation survives, showing **P** and **F** correctly as mobile and all other entities as stationary. **P** has been found moving to the left and **F** is approaching.

Figure 8.3 shows the viewpoint structure for the two interpretations active at $t = 0.5$ in this example. In the first part of Figure 8.3, node I-1 (on the left) is the root node and nodes I-2 to I-15 are the initial default hypotheses for the entities in the scene. Arrows indicate inheritance of facts between viewpoints, but not all existing links are shown in this diagram. Inner nodes of the graph correspond to *partial* interpretations that are combined to form *complete* interpretations in the second part of Figure 8.3. Node I-107 (on the very right) represents a single *complete* interpretation, formed by merging *partial* interpretations.

One purpose of experimenting with synthetic data is to verify the results of the reasoning process against the known 3-D structure and dynamics of the scene. Synthetic imagery also allows the simulation of various forms of motion which could not be obtained from available real data. For instance, none of the real image sequences that we had available contained an object crossing the path of the ALV (as does **P** in Figure 8.1).[1] After the reasoning process operated robustly on synthetic data, experiments were conducted on real ALV imagery. The results are described in the following section.

---

[1] An example of this situation in a real image sequence can be found in [12].

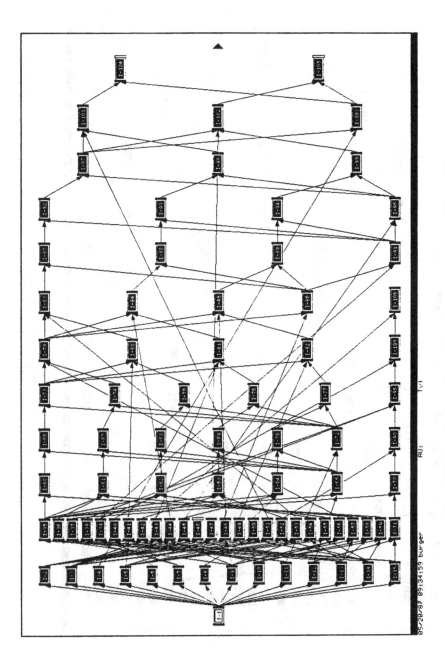

Figure 8.3 Snapshot from the evolving viewpoint structure of Figure 8.2.

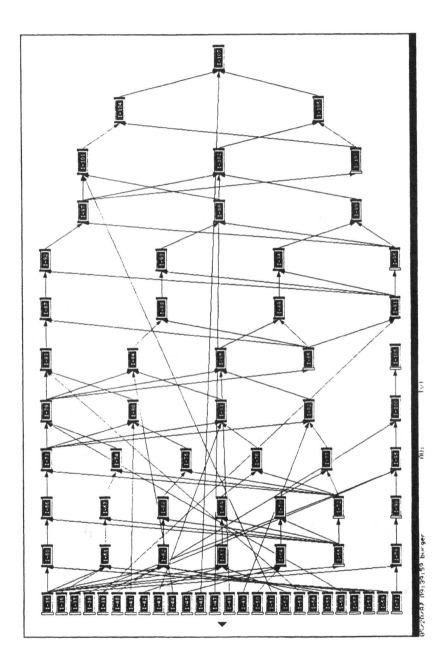

Figure 8.3 (*continued*).

8.2 REAL DATA

The data used for this experiment are part of an image sequence taken from the moving ALV (Figure 5.3). The binary edge-images of this sequence are shown in Figure 5.4. Points were tracked manually and the Fuzzy FOE algorithm (described in Chapter 5) was applied to each pair of frames. The results of this computation and the derotated displacement vectors are shown in Figure 5.5.

In addition to the rule base applied to the synthetic sequence in the previous section, the rate of divergence of image features away from the FOE was also used to relate the 3-D distances of individual entities. Consequently, it was possible to conclude more CLOSER-relationships than in the initial experiment.

Figures 8.4–8.18 show the complete scene interpretations starting at Frame 183 up to Frame 197 of that sequence. Interpretations are sorted by their number of stationary entities, i.e., *Interpretation-1* ranks higher than *Interpretation-2* if both exist and *Interpretation-1* has more stationary entities than *Interpretation-2*. During this run, the maximum number of concurrent interpretations was two. Entities are marked as *stationary* or *mobile* as in the previous example. Entities which carry no mark (just the label) are stationary and have not been found to be closer than any other entity in the scene. A square without a pointer in any direction means that this entity is considered *mobile*, but that the direction of movement could not be determined for the current frame.

The scene contains two moving objects, a car (entity **24**) which has passed the ALV and is moving away, and another vehicle (entity **33**) which is approaching the ALV on the same road. The second vehicle appears in Frame 185.

Frame 183

After the first pair of frames (Frames 182–183) two interpretations are created due to the movement of feature **24** which is located on a car that has passed the ALV and is now moving away in forward direction. Figure 8.4 shows two interpretations that were created in response to this 3-D motion:

Interpretation-1 (Figure 8.4a), which is the correct interpretation, "sees" only stationary entities, except point **24** which is considered mobile and appears to be moving upwards in the camera's coordinate frame. Quite a few CLOSER-relationships between the stationary scene entities have been established (indicated by straight lines between tokens). In contrast, point **24** is considered stationary in *Interpretation-2* (Figure 8.4*b*), while several other points (**8, 11**, ... **23**) appear to be moving *downwards*. Apparently, the movement of entity **24** was detected from relative motion with respect to this other set of points (see rule DIRECT-PAIR-MOTION in Section 6.4). Notice that in both interpretations, the two entities in the foreground (**4, 16**) are clearly seen to be closer than the rest of the scene.

In this case, *Interpretation-1* is ranked higher than *Interpretation-2* (and is thus considered more plausible) because of the larger number of stationary entities. At this point, however, there is not sufficient evidence to reject the other interpretation.

Frame 184

Interpretation-2 is eliminated after Frame 184 due to inconsistent divergence of the points considered moving downwards. In particular, some of the points that were considered moving downwards earlier (**8, 11**, ... **23**) were observed to be diverging away from the FOE (in upward direction), which is sufficient to reject this interpretation. Figure 8.5 shows the only remaining interpretation for Frame 184 with only point **24** being mobile. Notice that several entities that were close to the camera in the previous frame have meanwhile left the field of view and have also been eliminated from the model.

Frames 185–188

The single interpretation from Frame 184 is continued, because no object motion other than the one caused by entity **24** is observed in this period. The 3-D structure of the stationary part of the scene is continuously refined by adding new CLOSER-relationships between pairs of entities (Figures 8.6–8.9). The stationary entities **18, 19**, and **20** are apparently located at great distance for two reasons: *first*, because they are farther away than most other entities and, *second*, their relative motion is too small to infer anything about their relative depth. These points are in fact located at a distance of several hundred meters.

Frames 189–194

The single interpretation obtained in Frame 184 is continued till Frame 195. Entity **24** is the only entity considered mobile in this period (Figure 8.10– 8.15). Whenever the direction of motion could not be identified between a pair of frames, the mobile point is marked by a simple square. Notice the increased number of CLOSER-relationships established *across* the FOE-area caused by different rates of divergence of features away from the Fuzzy FOE.

Frame 195

After Frame 195, again two interpretations become feasible, this time caused by the movement of another car which is coming towards the ALV (entity **33**). In *Interpretation-1* (Figure 8.16*a*), point **33** is considered mobile and moving downwards. In *Interpretation-2* (Figure 8.16*b*), point **33** is stationary but several other points now appear to be in motion (**15, 39, 49, 50, 59, 60, 71**). These points were stationary in earlier, consistent interpretations, such that the current *Interpretation-2* is rather unlikely (although it cannot be rejected yet). Again, the correct *Interpretation-1* was ranked higher than *Interpretation-2* due to the larger number of stationary entities.

Frames 196–197

The two interpretations that originated in Frame 195 are pursued in parallel until Frame 197 (Figures 8.17–8.18). In Frame 196 (Figure 8.17), CLOSER-conflicts become visible in *Interpretation-2* between entities **33** and **36** as well as between **33** and **66**. The receding car (entity **24**) was not observed after Frame 196.

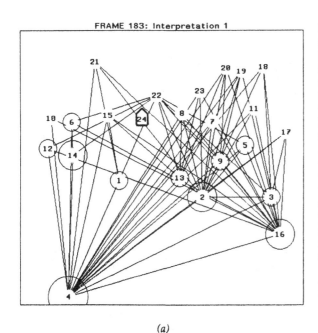

(a)

Figure 8.4 Interpretation of real ALV image sequence (Frame 183). After the first frame pair, two interpretations are created due to the movement of point **24**, which belongs to a car that has passed the ALV earlier and is now moving away. *Interpretation-1 (a)*, which is the correct interpretation, "sees" only stationary entities, except point **24** which is considered mobile and appears to be moving upwards in the camera's coordinate frame. In contrast, point **24** is stationary in *Interpretation-2 (b)*, while several other points (**8, 11, ... 23**) appear to be moving downwards.

(b)

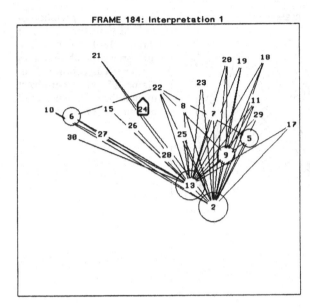

Figure 8.5 Interpretation of real ALV image sequence (Frame 184). *Interpretation-2* (from the previous figure) was found to be inconsistent and was discarded because some of the points that were considered moving downwards (8, 11, ... 23) were observed to be diverging away from the FOE. This leaves a single feasible interpretation for Frame 184 with only point 24 being mobile. Notice that several entities that were close to the camera in the previous frame have meanwhile left the field of view.

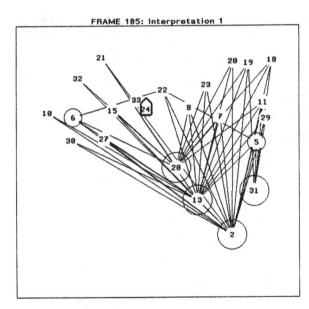

Figure 8.6 Interpretation of real ALV image sequence (Frame 185). The single interpretation obtained in Frame 184 is continued till Frame 195. No other object motion than the one caused by point 24 is observed. The 3-D model structure is continually refined by adding new CLOSER-relations. Notice that the entities 18, 19, and 20 must be located at great distance but nothing can be said about their relative position.

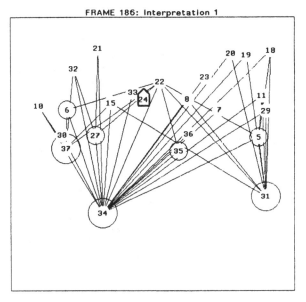

FRAME 186: Interpretation 1

Figure 8.7 Interpretation of real ALV image sequence (Frame 186). The single interpretation obtained in Frame 184 is continued till Frame 195. No other object motion than the one caused by point **24** is observed.

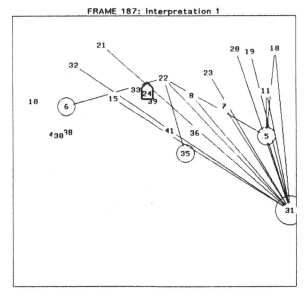

FRAME 187: Interpretation 1

Figure 8.8 Interpretation of real ALV image sequence (Frame 187).

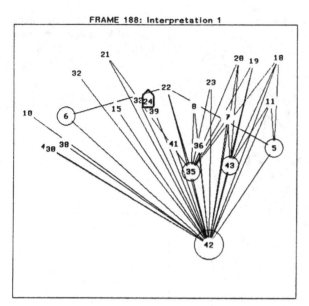

Figure 8.9 Interpretation of real ALV image sequence (Frame 188).

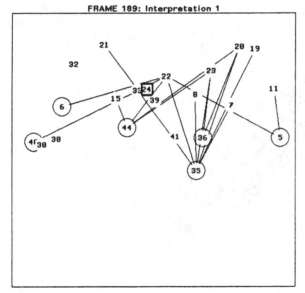

Figure 8.10 Interpretation of real ALV image sequence (Frame 189). Notice that, whenever the direction of motion could not be identified (as in this frame), the mobile point is marked by a simple square instead of a pointer.

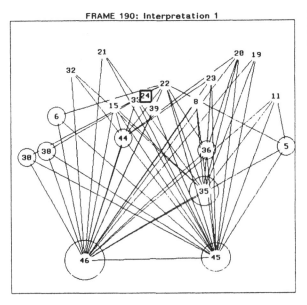

Figure 8.11 Interpretation of real ALV image sequence (Frame 190).

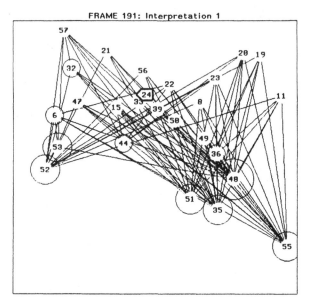

Figure 8.12 Interpretation of real ALV image sequence (Frame 191).

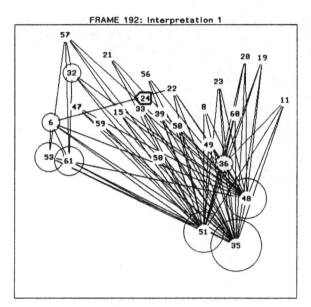

Figure 8.13 Interpretation of real ALV image sequence (Frame 192).

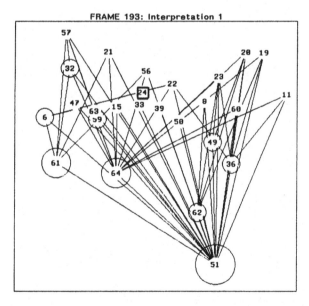

Figure 8.14 Interpretation of real ALV image sequence (Frame 193).

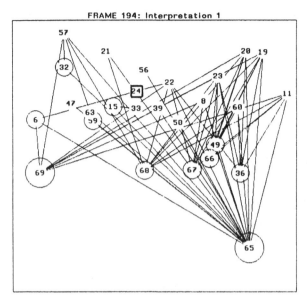

FRAME 194: Interpretation 1

Figure 8.15 Interpretation of real ALV image sequence (Frame 194).

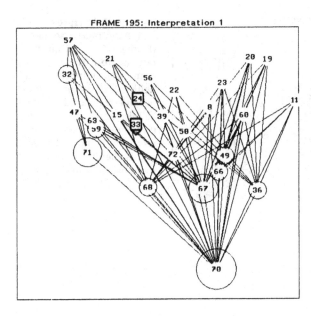

(a)

FRAME 195: Interpretation 2

(b)

Figure 8.16 Interpretation of real ALV image sequence (Frame 195). Again two different scene interpretations become feasible, this time caused by the movement of another car which is coming towards the ALV (point 33). In *Interpretation-1 (a)*, point 33 is mobile and moving downwards. In *Interpretation-2 (b)*, point 33 is stationary but several other points suddenly appear to be in motion.

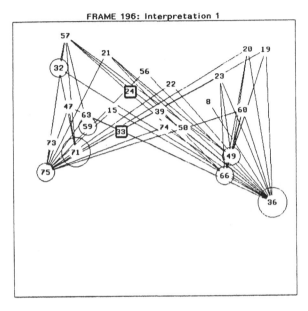

(a)

Figure 8.17 Interpretation of real ALV image sequence (Frame 196). Both interpretations from the previous frame are continued. So far, neither *Interpretation-1 (a)* nor *Interpretation-2 (b)* could be rejected. However, *Interpretation-1*, which is the correct one, is ranked higher and is therefore considered more plausible at this point.

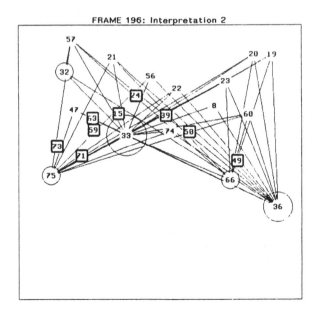

(b)

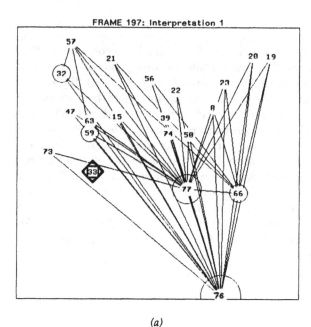

(a)

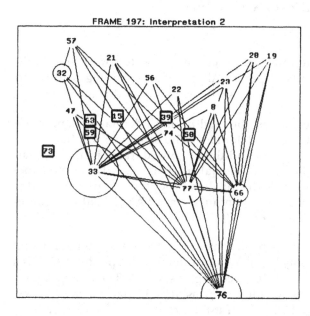

(b)

Figure 8.18 Interpretation of real ALV image sequence (Frame 197). The two interpretations from the previous frame are both continued. One of the mobile entities (point **24**) is not visible any more. In *Interpretation-1 (a)*, the mobile point **33** is found to be *approaching* the camera, indicated by a diamond-shaped marking. As **33** is located on a car coming towards the ALV, this interpretation is correct. The second interpretation *(b)* is already inconsistent but nevertheless shown here for illustration. The inconsistency arises mainly from the CLOSER-relationship between the (assumed stationary) entities **33** and **76**. In fact, **33** is hypothesized to be *nearer* than **76**, which is, however, much *lower* in the field of view. Thus, assuming that **33** is not a stationary point above the ground, there is conflicting evidence about the relative 3-D distances of **33** and **76**. This conflict is fatal for *Interpretation-2*. Only the correct *Interpretation-1* is continued.

Frame 197

The two interpretations from the previous frame are both continued. One of the mobile entities (point **24**) has meanwhile disappeared and was removed from the scene model. In *Interpretation-1* (Figure 8.18*a*), the mobile point **33** is found to be *approaching* the camera, indicated by a diamond-shaped marking. As **33** is located on a car coming towards the ALV, this interpretation is correct. The second interpretation (Figure 8.18*b*) is already inconsistent but still shown here to illustrate the process. The inconsistency of *Interpretation-2* arises mainly from the CLOSER-relationship between the entities **33** and **76** which are assumed to be stationary. In fact, **33** is hypothesized to be *nearer* than **76**, which is, however, much *lower* in the field of view. Thus, assuming that **33** is not a stationary point above the ground, there is conflicting evidence about the relative 3-D distances of **33** and **76**. This conflict is fatal for *Interpretation-2*. Only the correct *Interpretation-1* is continued. Entity **33** has been identified correctly as approaching the camera. All other entities of the scene are considered stationary.

8.3 IMPLEMENTATION ISSUES

With the exception of the low-level processing steps, this work was implemented on a *Symbolics* 3670 computer. All the figures showing synthesized or real images (except Figure 5.3) are screen copies from that machine.

The original image sequence was obtained from a video tape taken at the Martin-Marietta ALV test site in Colorado. The sequence was first transferred onto a magnetic video disc and then digitized frame by frame to an array of 512×512 pixels. From the original color signal, only the *Y*-component (luminance) was used in the further processing steps. The edge images shown in Figure 5.4 were computed on a VAX11/750 and subsequently transferred to the *Symbolics* 3670.

As mentioned earlier, the core of the reasoning process was implemented using the *Automated Reasoning Tool* (ART) by Inference Corp. [24], which is a widely used shell for building rule-based expert systems. The final program consisted of approximately 50 rules. A considerable amount of code was added in the form of supplementary LISP functions, which were used primarily for the FOE-computations, display-routines, and certain test functions to be used within rules.

Although speed was not a major concern in this project, care was taken to streamline the computation. Some of the measures found to be beneficial for improved performance were the following:

- The rules have been designed to have a most specialized condition pattern, such that only a small number of rules are affected whenever a new fact is asserted into the knowledge base.

- Whenever possible, rules were executed in blocks such that all other rules could be "switched off" by a special control pattern.

For our implementation, we have used an off-the-shelf expert system shell because it facilitates fast prototyping. Execution speed was only a secondary issue. Computing the FOE on the *Symbolics* took approximately 30 seconds for each frame on the ALV imagery. The execution time for the reasoning process (not including the FOE-computation) ranged from 2 to 3 minutes per frame for the experiments on the real images with about 25 points being tracked.

For the previous example, the image points used to create the initial displacement vectors were selected and tracked manually by a human operator. While it appears relatively easy to track distinct points in the original color or greyscale image sequence (Figure 5.3), this is much harder to do on the binary images shown earlier (Figure 5.4). As we have mentioned in Chapter 5, some points that are easy to track in original images do not stand out in the processed images, probably because some of the surrounding context information is lost. While manual feature tracking is, of course, not practical in any real application, we did not have a working algorithm available for this purpose at the time we made these experiments. In the meantime, algorithms for the automatic extraction of displacement vectors from real outdoor image sequences have become operational [12].

A special problem of driving sequences is that the appearance of interest points varies strongly as they get closer to the camera, which makes it difficult to track points reliably over extended periods. This is particularly true for points located on the ground. The disparity analysis algorithm developed in [12], therefore, uses range-dependent operators for the selection of interest points. In combination with our approach, this algorithm has been applied successfully to extended ALV image sequences, thus showing that automatic feature point selection and tracking is indeed practical. This also demonstrates that our reasoning approach works robustly even under the increased probability of erroneous displacement vectors (see [12] for details).

9

SUMMARY

The challenge of "understanding" image sequences obtained from a moving camera is that *stationary* objects in the 3-D environment are generally not static in the image and *mobile* objects in the scene do not always appear moving. Two-dimensional image-based techniques are clearly insufficient for this purpose, because the observed motion is directly related to the 3-D structure of the scene. In this paper, we presented the concepts and a new approach to scene understanding in dynamic environments. The viability of this approach has been demonstrated with a prototype system applied to real motion sequences.

Our approach departs from related work by following a strategy of *qualitative*, rather than quantitative, reasoning and modeling. While quantitative techniques have traditionally been dominant in computer vision, qualitative techniques are now receiving growing attention in this field [68, 73]. These methods hold the potential to replace expensive numerical computations and models by simpler reasoning about the relevant properties of the scene, relying upon less precise representations. They offer the possibility to make a tradeoff between the accuracy and the expressiveness of the results under certain conditions. This can result in a better overall usage of the limited computational resources. While this applies in particular to the higher levels of vision, we have shown that the qualitative approach can be used in *mediate–level* vision as well.

The numerical effort in our approach has been packed into the computation of the "Fuzzy FOE" (FFOE), which is an extension of the well-known FOE concept. This computation is a low–level process that is performed entirely in 2-D. The subsequent reasoning process evaluates the "derotated" displacement field with respect to the Fuzzy FOE and creates a qualitative description of the scene. One of the key features of this representation is that it contains not only a single interpretation but that *multiple scene interpretations* are pursued simultaneously. Unique to our approach is also the use of qualitative relations in 3-D to detect moving objects in the scene [14].

Since the interpretation of 2-D images is inherently ambiguous, the concurrent evaluation of different alternatives over a sequence of images provides a very powerful way of reasoning. Simultaneously evaluating a set of scene interpretations allows us to consider several alternatives and, depending upon the situation, select the appropriate one (e.g., the most "plausible" or the most "threatening" interpretation).

Experiments with this implementation were conducted with synthetic as well as real imagery. The results show that the tasks of detecting object motion, estimating camera motion, and determining the 3-D structure of the scene are accomplished successfully. While most of the development and experimentation was closely related to the Autonomous Land Vehicle, the technique is not restricted to this particular scenario but is applicable in a variety of scenarios for image understanding. As an example, the same system has been applied successfully to helicopter flight sequences without major modifications [12].

The availability of reliable displacement vectors is crucial to our approach. While we used manual point tracking for the examples shown here, recent results indicate that *automatic* feature selection and tracking have become practical [4, 12, 38]. In several experiments, the system described here was successfully tested on different ALV image sequences, each consisting of 20–80 frames, in a fully automatic mode [12]. Moreover, the use of inertial navigation sensors attached to the camera promises additional improvements for the computation of displacement vectors and passive ranging over purely visual techniques of motion analysis [62].

Naturally, most of the concepts presented here are far from being complete and future work could proceed in several directions. Among others, the focal points of further research could be, e.g.:

- Improving the representation and computation of the Fuzzy FOE to gain improvement in performance and allow more efficient inferencing.

- Qualitative grouping of scene entities based on common 3-D motion to represent rigid objects moving in space.

- Interactions between motion analysis and semantic object recognition, as indicated in Section 6.4.

- Streamlining the qualitative reasoning process, possibly by exploiting inherent parallelism.

Qualitative techniques will certainly not solve all the problems that exist in dynamic-scene and motion analysis, which remains one of the more difficult and challenging areas of vision research [14, 15]. However, being confident that the correspondence problem can be overcome, we believe that qualitative techniques, in general, do have the potential for solving practical problems in dynamic scene understanding. The work described here, as we hope, is an encouragement to proceed in that direction.

A

APPENDIX

A.1 GEOMETRIC CONSTRAINT METHOD FOR CAMERA MOTION

In the main part of this book (Chapter 4), we have described two approaches for computing the camera motion parameters from a set of displacement vectors. The two approaches (*FOE-from-Rotations* and *Rotations-from-FOE*) are actually attempts to solve the same continuous optimization problems in the space described by the four motion parameters u, v (the direction of heading, i.e., the location of the FOE) and θ, ϕ (the camera rotation angles). In this section, we describe a very different method for estimating these four parameters, which works by successively constraining the ranges of possible values using geometrical operations. The method is also based on the concept of intersecting displacement vectors. However, here we do not look for a *common* intersection point for all displacement vectors but use the weaker constraint that all vectors must *pass* through a given region.

The design of the ALV (and most other mobile robots) does not allow rapid changes in the direction of vehicle heading. Therefore, it can be assumed that the motion of the camera between two frames is constrained, such that the FOE can change its location only within a certain range. If the FOE is located in one frame, the FOE in the subsequent frame must lie within a certain image region around the previous FOE location.

Testing FOE–Feasibility

We now assume that some image region **S** contains the current FOE, but the exact location of the FOE and the camera rotations are unknown. We want to verify if the region **S** really contains the FOE (i.e., if it is *FOE-feasible*), without having to compute the exact location of the FOE.

If the region **S** is *FOE-feasible*, then every displacement vector, after being de-rotated by some rotation angles θ, ϕ, must intersect the region **S**. In other words, an image region **S** is defined to be *FOE-feasible* if such rotation angles θ, ϕ exist. The problem is to show that θ and ϕ exist without computing their exact values. To accomplish this, we try to prove the contrary, i.e., no such rotations exist and the region **S** cannot possibly contain the FOE.

If a particular displacement vector $\mathbf{u}_i = \mathbf{x}_i \rightarrow \mathbf{x}_i'$ should be derotated in order to intersect a given region **S**, this can be accomplished in general by an infinite number of possible combinations of rotation angles θ and ϕ. We denote this *range* of possible derotation angles for the displacement vector $\mathbf{u}_i$ by $\mathbf{R}_i(\mathbf{S}) = \{(\theta' \, \phi')\}$. A necessary condition for the existence of rotation angles $(\theta \, \phi)$ that would cause *all* displacement vectors to intersect **S**, is that the intersection of all $\mathbf{R}_i(\mathbf{S})$ must be non-empty (for all displacement vectors $\mathbf{u}_i$, $i = 1...N$). Unfortunately, this does not guarantee that **S** really contains the FOE. However, if the intersection of all $\mathbf{R}_i(\mathbf{S})$ *is* empty, then **S** is *not FOE-feasible*.

Figure A.1 illustrates the situation. In Figure A.1*a*, the hypothesized image region **S** (thought to contain the FOE) is marked by a rectangle that is centered at the location of the previous FOE. Three displacement vectors are shown (**P1**→**P1'**, **P2**→**P2'**, **P3**→**P3'**) which are the result of camera translation and rotation. Also shown are the translational displacement vectors (**P1**→**Q1**, **P2**→**Q2**, **P3**→**Q3**) and the true FOE (marked by a small circle). In Figure A.1*b*, the range of possible rotations $\mathbf{R}_i(\mathbf{S})$ is represented by a closed polygon, called the *rotation polygon*, in the two-dimensional *rotation space*. The two coordinate axes of the rotation space, *theta* and *phi*, correspond to the amount of camera rotations (θ, ϕ) about the *Y*-axis and the *X*-axis respectively. The initial *rotation polygon* (a square) covers a range of $\pm 10°$ in both directions.

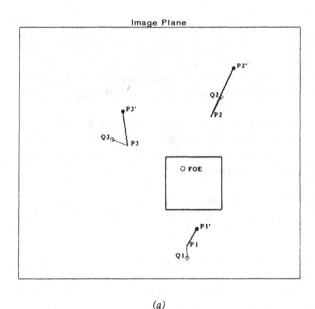

(a)

Figure A.1 Image space and rotation space. The displacement field in the image plane *(a)* contains three displacement vectors (**P1→P1′**, **P2→P2′**, **P3→P3′**). The previous FOE was observed at the center of the square. This is the region of search for the current FOE. The translational displacement components (**P1→Q1**, **P2→Q2**, **P3→Q3**) and the current location of the FOE are unknown but marked in this picture. The initial range of possible camera rotations is ±10° in either direction, which corresponds to the inner region of the square in the rotation space *(b)*, i.e., initially any combination of rotations (θ, ϕ) within that square is considered possible.

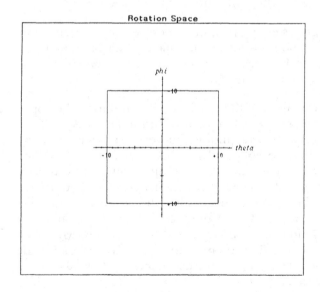

(b)

Algorithm

Starting from the initial range of possible rotations, we compute the intersection of all $R_i(S)$ by successively constraining the rotation polygon. Given the hypothesized FOE-region S and the current rotation polygon R, the following steps are executed for each displacement vector $P_j \rightarrow P_j'$ (Figure A.2):

(1)   For each vertex of the current rotation polygon, $\omega_j = (\theta_j, \phi_j) \in R$, apply the inverse rotation mappings $\rho_{\phi,\theta}^{-1}$ (Equation 4.4 in Chapter 4) to the end point P_j' of the displacement vector. This yields a set of points p_j in the image space (Step 1 in Figure A.2a).

(2)   Link the points p_j to form a closed polygon in the image space. The resulting polygon is similar to the rotation polygon R but distorted by the nonlinear rotation mapping (Figure A.2a).

(3)   Intersect the polygon in the image with the sector formed by the starting point P of the displacement vector and the two tangents onto the FOE-region S (Step 2 in Figure A.2b). The result is a new (possibly empty) polygon in the image plane.

(4)   Map the new polygon from the image plane back to rotation space (using Equation 3.14 of Section 3). This is Step 3 in Figure A.2b.

(5)   If the rotation polygon is empty (i.e., number of vertices is zero) then no camera rotation is possible that would make all displacement vectors intersect the given FOE-region S, i.e., S is *not FOE-feasible*. Otherwise the region S may contain the FOE.

Figures A.3*a–f* show the evolution of the rotation polygon during the application of this process to the three displacement vectors in Figure A.1. As an effect of perspective imaging, the straight lines between vertices in the rotation polygon correspond to hyperbolic paths in the image space (Equations 3.11 and 3.12 in Chapter 3). They are, however, approximated as straight lines in the image space in order to simplify the intersection operation in Step 3. The dotted lines in the image space show the *real* mapping of the rotation polygon onto the image. Unless the focal length of the camera is extremely short, the deviations from straight lines are small enough to be ignored. Figure A.3*f* shows the final rotation polygon after evaluating all three displacement vectors. The amount of true camera rotation ($\theta = -2.0°$, $\phi = 5.0°$) is marked with a small circle.

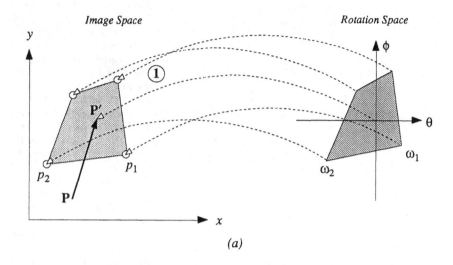

(a)

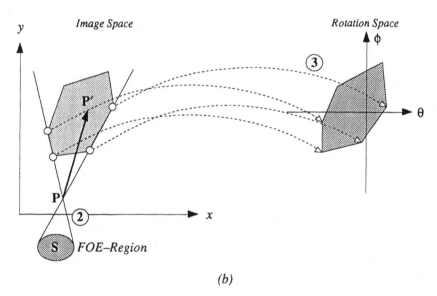

(b)

Figure A.2 Successively constraining the range of possible camera rotations. *(a)* The camera rotations corresponding to the vertices of the current rotation polygon are applied to every displacement vector (**P**→**P′**) in Step 1. This yields another polygon in the image space with vertices p_i. *(b)* The image polygon is intersected with the sector bounded by the tangents from **P** to the FOE–region **S** (Step 2). Rotations that would bring the end point of the displacement vector *outside* this sector are not feasible. The new vertices on the image space polygon are mapped back into rotation space (Step 3).

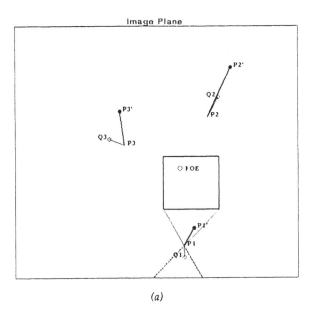

(a)

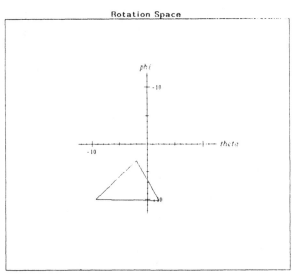

(b)

Figure A.3 Evolution of the rotation polygon. The FOE is assumed to be located somewhere inside the square image region shown in *(a)*. The goal is now to "rotate" the first displacement vector **P1→P1′**, such that it intersects with the FOE region. To do this, any rotated end point **P1″** = $r_{\phi'\theta'}$(**P1′**) must lie within the sector bounded by the two tangents from the FOE region that pass through **P1**. **Q1** (the end point of the translational displacement vector) is located in this sector. Mapping this sector from image space to rotation space *(b)* yields the modified rotation polygon (a triangle in this case).

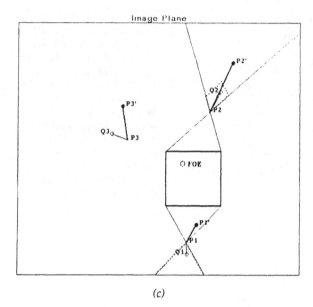

(c)

Figure A.3 (*continued*) Now displacement vector **P2→P2′** is evaluated. First *(c)*, the rotations corresponding to the vertices of the previous rotation polygon (Figure A.3*b*) are applied to the end point **P2′**, creating a similar polygon in the image space. This polygon is intersected with the sector bounded by the two tangents from the FOE region that pass through **P2** (**Q2** is located in this region). Mapping the resulting polygon from image space to rotation space *(d)* yields the modified rotation polygon (now with 4 vertices).

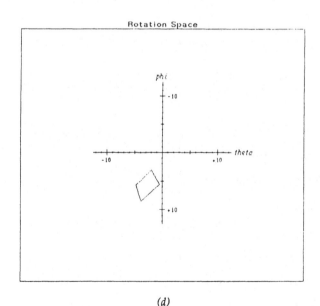

(d)

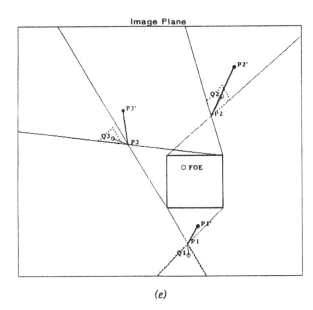

(e)

Figure A.3 (*continued*) The rotation polygon (*f*) and its image-space counterpart (*e*) after evaluating displacement vectors **P1→P1′**, **P2→P2′**, and **P3→P3′**. The value of the true camera rotation is marked in (*f*) with a small circle (pointed out by an arrow). Since the range of possible rotations is not empty, the square region is *FOE-feasible* with respect to the 3 displacement vectors.

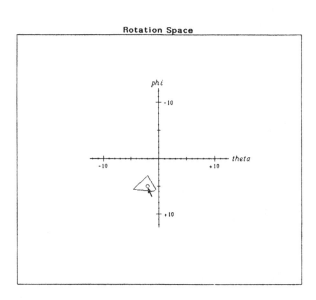

(*f*)

Of course, increasing the number of displacement vectors improves the rotation estimate. In practice, the amount of camera rotation can be constrained to a range of below 1° in both directions. It is interesting (although not surprising) that rotations can be estimated more accurately when the displacement vectors are short, i.e., when the amount of camera translation is small. This is in contrast to estimating camera translation, which is easier to compute with long displacement vectors.

Searching for the FOE

The situation when the rotation polygon becomes *empty* requires some additional considerations. As mentioned earlier, in such a case no camera rotation is possible that would make all displacement vectors pass through the given FOE-region. This could indicate one of the two alternatives:

(a)   At least one of the displacement vectors belongs to a moving object.

(b)   The given FOE-region does not contain the actual location of the FOE, i.e., the region is *not FOE-feasible.*

The latter case is of particular importance. If a region can be determined *not* to contain the FOE, then the FOE must necessarily lie outside this region. Therefore, the above method can not only be used to estimate the amount of camera rotation, but also to search for the location of the FOE.

Unfortunately, if the rotation polygon does not become empty, this does *not* imply that the FOE is actually inside the given region. It only means that all displacement vectors would *pass through* this region (for the allowed range of rotations), not that they have a common *intersection* inside this region. However, if not all vectors pass through a certain region, then this region cannot possibly contain the FOE. The following algorithm **Min-Feasible** searches a given region for the FOE by splitting it recursively into smaller pieces in a coarse-to-fine fashion:

Min-Feasible (region, min-size, disp-vectors):
    if Size (region) < min-size *then*
        return (region);   */* region may contain the FOE */*
    else
        if FOE-Feasible (region, disp-vectors) *then*
          (subregion-1,..., subregion-N) := *split-region* (region)
            return (*union* (
                Min-Feasible (subregion-1, min-size, disp-vectors),
                Min-Feasible (subregion-2, min-size, disp-vectors),

                ....
                Min-Feasible (subregion-N, min-size, disp-vectors)));
        else return (nil);   */* region does not contain the FOE */*

This algorithm searches for the smallest feasible FOE-region by systematically discarding subregions from further consideration. For the case that the shape of the original region is a square, subregions can be obtained by splitting the region into four sub-squares of equal size. The above version of the algorithm *Min-Feasible* performs a depth-first search down to the smallest subregion (limited by the parameter min-size). This is neither the most elegant nor the most efficient approach. The algorithm can be significantly improved by applying a more sophisticated strategy, e.g., by trying to discard subregions around the perimeter first before examining the interior of a region.

Two major problems were encountered with this method. *First,* the algorithm is computationally expensive, since the process of computing feasible rotations must be repeated for every subregion. *Second,* a small region is more likely to be discarded than a larger one. However, when the size of the region becomes too small, errors induced by noise, distortion, or point-tracking may prohibit displacement vectors from passing through a region which actually contains the FOE.

While this particular method was not employed in the final DRIVE implementation, its concept is directly related to the "Fuzzy FOE" (see Chapter 5). The Fuzzy FOE algorithm in Chapter 5 searches for the "best" FOE first and then grows a compact region with certain properties around that location, thus working "from inside out". The method presented in this section works in the opposite direction by first assuming a large region of possible FOE locations that is successively constrained. Therefore, the method never deals with individual FOE locations but always operates on FOE *regions* and corresponding *ranges* of possible camera rotations.

A.2 ESTIMATING ABSOLUTE VELOCITY

If a set of stationary 3-D points X_i is observed by a moving camera, then the translation in the Z-direction (ΔZ) is the same for every point, i.e.,

$$Z_i - Z_i' = Z_j - Z_j' = \Delta Z \quad \forall i,j. \tag{A.1}$$

Therefore, the depth Z_i of every point is proportional to the observed amount of divergence of its image away from the FOE x_f,

$$Z_i \propto \frac{\left\| x_i' - x_f \right\|}{\left\| x_i' - x_i \right\|} \tag{A.2}$$

which renders the relative 3-D structure for the set of observed points x_i.

Given a translational displacement field and the location of the FOE, the 3-D layout of the scene can be obtained up to a common scale factor (Equation 3.20 in Section 3). This scale factor (and consequently the velocity of the vehicle) can be determined if the 3-D position of one point in space is known. Furthermore, it can be shown that it is sufficient to know only *one* coordinate value of a point in space to reconstruct its position in space from its location in the image [25, 31].

Since the ALV travels on a fairly flat surface, the road can be approximated as a plane which lies parallel to the vehicle's direction of translation (Figure A.4). This approximation holds at least for a good part of the road in the field of view of the camera. Because the absolute height of the camera above the ground is constant and known, it is possible to estimate the positions of points on the road surface with respect to the vehicle in *absolute* terms. From the changing distances between these points and the camera, the actual advancement and speed of the vehicle can be determined.

First, a new coordinate system is introduced which has its origin in the lens center of the camera. The Z-axis of the new system passes through the FOE in the image plane and points, therefore, in the direction of translation. The original camera-centered coordinate system ($X\ Y\ Z$) is transformed into the new frame ($X'\ Y'\ Z'$) merely by applying horizontal and vertical rotations until the Z-axis lines-up with the FOE.

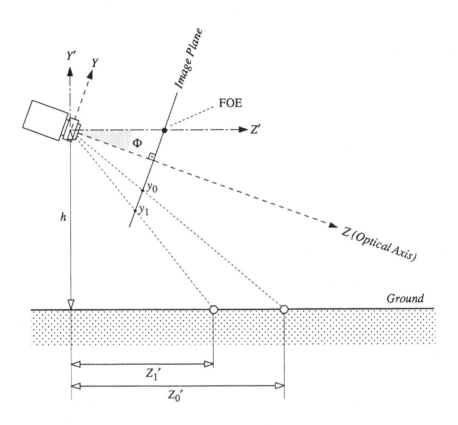

Figure A.4 Side view of a camera moving over a flat surface. The camera advances in direction Z', such that a 3-D point on the ground moves relative to the camera from Z_0' to Z_1'. The inclination angle Φ can be determined from the location of the FOE in the image. The absolute height of the camera above the ground (h) is assumed to be known.

The horizontal and vertical orientation in terms of *pan* and *tilt* are obtained by "rotating" the FOE $(x_f y_f)$ into the center of the image $(0\ 0)$ using Equation 3.14 from Section 3:

$$\theta_f = -\tan^{-1}\frac{x_f}{f} \tag{A.3}$$

$$\phi_f = -\tan^{-1}\left(f^2\frac{y_f}{(f^2+x_f^2)f^2 - x_f^2 y_f^2}\right). \tag{A.4}$$

The two angles θ_f and ϕ_f represent the orientation of the camera in 3-D with respect to the new coordinate system. This again allows to determine the 3-D orientation of the projecting rays passing through image points by use of the inverse perspective transformation. A 3-D point X in the environment whose image $x = (x\ y)$ is given, lies on a straight line in space defined by

$$X = \begin{bmatrix} X \\ Y \\ Z \end{bmatrix} = \kappa \begin{bmatrix} \cos\phi_f & \sin\theta_f\sin\phi_f & -\sin\theta_f\cos\phi_f \\ 0 & \cos\theta_f & \sin\phi_f \\ \sin\phi_f & -\cos\theta_f\sin\phi_f & \cos\theta_f\cos\phi_f \end{bmatrix} \cdot \begin{bmatrix} x \\ y \\ f \end{bmatrix} \tag{A.5}$$

for $\kappa \in \mathbb{R}$. For points on the road surface, the Y-coordinate is $-h$, where h is the height of the camera above ground. Therefore, the value of κ for a road surface point observed at $(x_s\ y_s)$ can be estimated as

$$\kappa_s = \frac{-h}{y_s\cos\phi_f + f\sin\phi_f} \tag{A.6}$$

and its 3-D distance is found by replacing κ in (A.5) as

$$Z_s = -h\ \frac{x_s\sin\theta_f - y_s\cos\theta_f\sin\phi_f - f\cos\theta_f\cos\phi_f}{y_f\cos\phi_f + f\sin\phi_f} \tag{A.7}$$

If a point on the ground is observed at two instances of time, x_s at time t and x_s' at t', the resulting distances from the vehicle Z_s at t and Z_s' at t' yield the amount of advancement $\Delta Z(t,t')$ and velocity $V(t,t')$ during this period as

$$\Delta Z(t, t') = Z_s - Z_s' \tag{A.8}$$

and

$$V(t, t') = \frac{\Delta Z(t, t')}{t - t'} \ . \tag{A.9}$$

Of course, image noise and tracking errors may have a large impact on the the quality of this velocity estimate. Therefore, the longest available displacement vectors from points on the road surface are preferred for this measurement, i.e., points that are relatively close to the vehicle. Also, the ground surface is never perfectly flat. Obviously, a more robust velocity estimate can be obtained by combining the measurements from several displacement vectors. Given a set of suitable displacement vectors, we compute the new estimate ΔZ^* as the weighed average of the individual measurements ΔZ_i :

$$\Delta Z^*(t, t') = \frac{\sum \left(\left\| \mathbf{x}_i - \mathbf{x}_i \right\| \Delta Z_i(t, t') \right)}{\sum \left\| \mathbf{x}_i - \mathbf{x}_i \right\|} \ . \tag{A.10}$$

Notice that the length of each displacement vector is used as the weight for its contribution to the final estimate. The combined velocity estimate V^* then becomes

$$V^*(t, t') = \frac{\Delta Z^*(t, t')}{t - t'} \ . \tag{A.11}$$

The values shown in Section 5 (Figure 5.5) for the absolute advancement between each frame pair were computed using (A.10).

REFERENCES

1. Adiv G., "Determining Three-Dimensional Motion and Structure from Optical Flow Generated by Several Moving Objects," *IEEE Trans. on Pattern Analysis and Machine Intelligence* 7 (4), pp. 384–401 (1985).

2. Aloimonos J., and Brown C.M., "Direct Processing of Curvilinear Sensor Motion from a Sequence of Perspective Images," Proc. IEEE Workshop on Computer Vision, pp. 72–77 (April 1984).

3. Anandan P., "Computing Dense Displacement Fields With Confidence Measures in Scenes Containing Occlusion," Proc. SPIE Intelligent Robots and Computer Vision 521, pp. 184–194 (1984).

4. Anandan P., "A Computational Framework and an Algorithm for the Measurement of Visual Motion," *International Journal of Computer Vision* 2, pp. 283–310 (1989).

5. Bandopadhay A., Chandra B., and Ballard D.N., "Egomotion Using Active Vision," Proc. IEEE Conf. on Computer Vision and Pattern Recognition CVPR'86, pp. 498–503 (1986).

6. Barnard S.T., and Thompson W.B., "Disparity Analysis of Images," *IEEE Trans. on Pattern Analysis and Machine Intelligence* 2 (4), pp. 333–340 (July 1980).

7. Bhanu B., and Burger W., "DRIVE – Dynamic Reasoning from Integrated Visual Evidence," Proc. DARPA Image Understanding Workshop, Los Angeles, pp. 581–588, Morgan Kaufmann Publishers (Feb. 1987).

8. Bhanu B., and Burger W., *DRIVE – Dynamic Reasoning from Integrated Visual Evidence*, DARPA Contract DACA 76-86-C-0017 Project Report,

U.S. Army Engineer Topographic Laboratories, Fort Belvoir, VA (June 1987).

9. Bhanu B., Nasr H., Symosek P., Burger W., Gluch M., and Panda D., "Dynamics for an Autonomous Land Vehicle," Proc. Conf. Association for Unmanned Vehicle Systems, Washington D.C. (July 1987).

10. Bhanu B., and Burger W., "Approximation of Displacement Fields Using Wavefront Region Growing", *Computer Vision, Graphics and Image Processing* **41**, pp. 306–322 (March 1988).

11. Bhanu B., and Burger W., "Qualitative Motion Detection and Tracking of Targets from a Mobile Platform," Proc. DARPA Image Understanding Workshop, Cambridge Mass., Morgan Kaufmann Publishers, pp. 289–318 (April 1988).

12. Bhanu B., Symosek P., Ming J., Burger W., Nasr H., and Kim J., "Qualitative Target Motion Detection and Tracking," Proc. DARPA Image Understanding Workshop, Palo Alto CA, Morgan Kaufmann Publishers, pp. 370–397 (May 1989).

13. Bhanu B., Burger W., "A Qualitative Approach to Dynamic Scene Understanding," *Computer Vision, Graphics, and Image Processing: Image Understanding* **54** (2), pp. 184–205 (September 1991).

14. Bhanu B., Nevatia R., Riseman E.M., "Dynamic Scene and Motion Analysis Using Passive Sensors I: Qualitative Approach," *IEEE Expert* **7** (1), pp. 45–52 (February 1992).

15. Bhanu B., Nevatia R., Riseman E.M., "Dynamic Scene and Motion Analysis Using Passive Sensors II: Displacement Field and Feature-Based Approaches," *IEEE Expert* **7** (1), pp. 53–64 (February 1992).

16. Bharwani S., Riseman E., and Hanson A., "Refinement Of Environmental Depth Maps Over Multiple Frames," Proc. IEEE Workshop on Motion: Representation and Analysis, pp. 73–80 (May 1986).

17. Bolles R.C., and Baker H.H., "Epipolar–Plane Analysis: A Technique for Analyzing Motion Sequences," Proc. IEEE Workshop on Computer Vision: Representation and Control, pp. 168–178 (October 1985).

18. Broida T.J., and Chellappa R., "Estimation of Object Motion Parameters from Noisy Images," *IEEE Trans. on Pattern Analysis and Machine Intelligence*, **8** (1), pp. 90–99 (January 1986).

19. Bruss A.R., and Horn B.K.P., "Passive Navigation," *Computer Vision, Graphics and Image Processing* **21**, pp. 3–20 (1983).

20. Burger W., and Bhanu B., "Qualitative Motion Understanding," Proc. Intl. Joint Conf. on Artificial Intelligence IJCAI'87, Milan, Morgan Kaufmann Publishers, pp. 819–821 (Aug. 1987).

21. Burger W., and Bhanu B., "Dynamic Scene Understanding for Autonomous Mobile Robots," Proc. IEEE Conf. Computer Vision and Pattern Recognition CVPR'88, Ann Arbor, Michigan, pp. 736–741 (June 1988).

22. Burger W., Bhanu B., "Estimating 3-D Egomotion from Perspective Image Sequences," *IEEE Trans. on Pattern Analysis and Machine Intelligence*, **12** (11), pp. 1040–1058 (November 1990).

23. Burger W., Bhanu B., "Qualitative Understanding of Scene Dynamics for Mobile Robots," *The International Journal of Robotics Research* **9** (6), pp. 74–90 (December 1990).

24. Clayton B.D., *ART Programming Manual*, Inference Corp., Los Angeles (1985).

25. Duda R., and Hart P., *Pattern Classification and Scene Analysis*, Wiley, New York (1973).

26. Dutta R., Manmatha R., Riseman E.M., and Snyder M.A., "Issues in Extracting Motion Parameters and Depth from Approximate Translational Motion," Proc. DARPA Image Understanding Workshop, Cambridge Mass., Morgan Kaufmann Publishers, pp. 945–960 (April 1988).

27. Fang J.Q., and Huang T.S., "Some Experiments on Estimating the 3–D Motion Parameters of a Rigid Body from Two Consecutive Image Frames," *IEEE Trans. on Pattern Analysis and Machine Intelligence* **6** (5), pp. 545–554 (September 1984).

28. Faugeras O.D., Lustman F., and Toscani G., "Motion and Structure from Point and Line Matches," Proc. Intl. Conf. on Computer Vision ICCV'87, pp. 25–34 (1987).

29. Freeman J., "The Modelling of Spatial Relations," *Computer Graphics and Image Processing* 4, pp. 156–171 (1975).

30. Hagert G., "What's in a Mental Model? On Conceptual Models in Reasoning with Spatial Descriptions," Proc. Intl. Joint Conf. on Artificial Intelligence IJCAI'85, Morgan Kaufmann Publishers, pp. 274–277 (1985).

31. Haralick R.M., "Using Perspective Transformations in Scene Analysis," *Computer Graphics and Image Processing* 13, pp. 191–221 (1980).

32. Hildreth E.C., *The Measurement of Visual Motion*, MIT Press, Cambridge, Mass. (1984).

33. Hildreth E.C., and Grzywacz N.M., "The Incremental Recovery of Structure from Motion: Position vs. Velocity Based Formulations, Proc. IEEE Workshop on Motion: Representation and Analysis, pp. 137–143 (May 1986).

34. Horn B.K.P., and Schunck B.G., "Determining Optical Flow," *Artificial Intelligence* 17, pp. 185–203 (Aug. 1981).

35. Jain R., "Direct Computation of the Focus of Expansion," *IEEE Trans. on Pattern Analysis and Machine Intelligence* 5 (1), pp. 58–64 (Jan. 1983).

36. Jain R., Barlett S.L., and O'Brien N., "Motion Stereo Using Ego–Motion Complex Logarithmic Mapping," Proc. IEEE Conf. Computer Vision and Pattern Recognition CVPR'86, pp. 188–193 (1986).

37. Jerian C., and Jain R., "Determining Motion Parameters for Scenes with Translation and Rotation," *IEEE Trans. on Pattern Analysis and Machine Intelligence* 6 (4), pp. 523–530 (July 1984).

38. Kim J., and Bhanu B., "Motion Disparity Analysis Using Adaptive Windows," Technical Report 87SRC38, Honeywell Systems & Research Center, Minneapolis MN (June 1987).

39. Kories R., and Zimmermann G., "A Versatile Method for the Estimation of Displacement Vector Fields from Image Sequences," Proc. IEEE Workshop on Motion: Representation and Analysis, pp. 101–106 (May 1986).

40. Kuipers B., "Qualitative Simulation," *Artificial Intelligence* **29**, pp. 289–338 (1986).

41. Lawton D.T., "Processing Translational Motion Sequences," *Computer Vision, Graphics and Image Processing* **22**, pp. 114–116 (1983).

42. Lee D.N., "The Optic Flow Field: The Foundation of Vision," *Phil. Trans. R. Soc. Lond. B* **290**, pp. 169–179 (1980).

43. Longuet–Higgins H.C., and Prazdny K., "The Interpretation of a Moving Retinal Image," *Proc. R. Soc. London B* 208, pp. 385–397 (1980).

44. Longuet–Higgins H.C., "A Computer Algorithm for Reconstructing a Scene from Two Projections," *Nature* **293**, pp. 133–135 (Sept. 1981).

45. Marimont D.H., "Projective Duality and the Analysis of Image Sequences," Proc. IEEE Workshop on Motion: Representation and Analysis, pp. 7–14 (May 1986).

46. Matthies L., Szeliski R., and Kanade T., "Kalman Filter-Based Algorithms for Estimating Depth from Image Sequences," Proc. DARPA Image Understanding Workshop, Cambridge Mass., Morgan Kaufmann Publishers, pp. 199–213 (April 1988).

47. Mitiche A., Seida S., and Aggarwal J.K., "Determining Position and Displacement in Space from Images," Proc. IEEE Conf. on Computer Vision and Pattern Recognition CVPR'85, San Francisco, pp. 504–509 (June 1985).

48. Moravec H.P. , "Towards Automatic Visual Obstacle Avoidance," Proc. Intl. Joint Conf. on Artificial Intelligence IJCAI'77, pp. 584 (Aug. 1977).

49. Nagel H.-H., "Representation of Moving Objects Based on Visual Observations," *IEEE Computer* pp. 29–39 (August 1981).

50. Nagel H.-H., "Displacement Vectors Derived from Second–Order Intensity Variations in Image Sequences," *Computer Vision, Graphics, and Image Processing* **21**, pp. 85–117 (1983).

51. Nagel H.-H., "Image Sequences – Ten (octal) Years – From Phenomenology towards a Theoretical Foundation, Proc. Intl. Conf. on Pattern Recognition ICPR'86, Paris, pp. 1174–1185 (1986).

52. Nii H.P., "Blackboard Systems: The Blackboard Model of Problem Solving and the Evolution of Blackboard Architectures," *The AI Magazine*, pp. 38–53 (August 1986).

53. Nii H.P., "Blackboard Application Systems, Blackboard System from a Knowledge Engineering Perspective," *The AI Magazine*, pp. 82–106 (August 1986).

54. Perrone J.A., "Anisotropic Responses to Motion Toward and Away from the Eye," *Perception & Psychophysics* **39** (1), pp. 1–8 (1986).

55. Prager J.M., and Arbib M.A., "Computing the Optic Flow: The MATCH Algorithm and Prediction," *Computer Vision, Graphics and Image Processing* **24**, pp. 271–304 (1983).

56. Prazdny K., "Determining the Instantaneous Direction of Motion from Optical Flow Generated by a Curvilinear Moving Observer," *Computer Graphics and Image Processing* **17**, pp. 238–248 (1981).

57. Prazdny K., "On the Information in Optical Flows," *Computer Vision, Graphics, and Image Processing* **22**, pp. 239–259 (1983).

58. Regan D., Beverly K., and Cynader M., "The Visual Perception of Motion in Depth," *Scientific American*, pp. 136–151 (July 1979).

59. Rieger J.H., "Information in Optical Flows Induced by Curved Paths of Observation," *J. Opt. Soc. Am.* **73** (3), pp. 339–344 (March 1983).

60. Rieger J.H., and Lawton D.T., "Processing Differential Image Motion," *J. Opt. Soc. Am. A* **2** (2), pp. 354–360 (Feb. 1985).

61. Roach J.W., and Aggarwal J.K., "Determining the Movements of Objects from a Sequence of Images," *IEEE Trans. on Pattern Analysis and Machine Intelligence* 2 (6), pp. 554–562 (1980).

62. Roberts B., and Bhanu B., "Inertial Navigation Sensor Integrated Motion Analysis for Autonomous Vehicle Navigation," Proc. DARPA Image Understanding Workshop, Pittsburgh, PA, Morgan Kaufmann Publishers, pp. 364–375 (Sept. 1990).

63. Schunck B.G., "Image Flow: Fundamentals and Future Research," Proc. IEEE Conf. on Computer Vision and Pattern Recognition CVPR'85, pp. 560–571 (1985).

64. Sethi I.K., and Jain R., "Finding Trajectories of Feature Points in a Monocular Image Sequence," *IEEE Trans. on Pattern Analysis and Machine Intelligence* 9 (1), pp. 56–73 (Jan. 1987).

65. Shariat H., and Price K.E., "How to Use More Than Two Frames to Estimate Motion," Proc. IEEE Workshop on Motion: Representation and Analysis, pp. 119–124 (May 1986).

66. Sridhar B., Cheng V.H.L., and Phatak A.V., "Kalman Filter Based Range Estimation for Autonomous Navigation Using Imaging Sensors," Proc. 11th IFAC Symposium on Automatic Control in Aerospace, Tsukuba, Japan (July 1989).

67. Thompson W.B., and Barnard S.T., "Lower–Level Estimation and Interpretation of Visual Motion," *IEEE Computer* 14 (8), pp. 20–27 (August 1981).

68. Thompson W.B., and Kearney J.K., "Inexact Vision," Proc. IEEE Workshop on Motion: Representation and Analysis, pp. 15–21 (1986).

69. Tsai R.Y., and Huang T.S., "Uniqueness and Estimation of Three–Dimensional Motion Parameters of Rigid Objects with Curved Surfaces," *IEEE Trans. on Pattern Analysis and Machine Intelligence* 6 (1), pp. 13–27 (Jan. 1984).

70. Ullman S., *The Interpretation of Visual Motion*, MIT Press, Cambridge, Mass. (1979).

71. Ullman S., "Computational Studies in the Interpretation of Structure and Motion: Summary and Extensions," MIT A.I. Memo No. 706 (March 1983).

72. Ullman S., "Maximizing Rigidity: The Incremental Recovery of 3–D Structure from Rigid and Rubbery Motion," MIT A.I. Memo No. 721 (June 1983).

73. Verri A., and Poggio T., "Qualitative Information in the Optical Flow," Proc. DARPA Image Understanding Workshop, Los Angeles, pp. 825–834 (Feb. 1987).

74. Warren W.H., and Hannon D.J., "Direction of Self-Motion Is Perceived from Optical Flow," *Nature* **336**, pp. 162–163 (1988).

75. Warren W.H., Morris M.W., and Kalish M., "Perception of Translation Heading from Optical Flow," *Journal of Experimental Psychology: Human Perception and Performance* **14**, pp. 646–660 (1988).

76. Warren W.H., Mestre D.R., Blackwell A.W., and Morris M.W., "Perception of Circular Heading from Optical Flow," *Journal of Experimental Psychology: Human Perception and Performance* **17**, pp. 28–43 (1991).

77. Weng J., Huang T.S., and Ahuja N., "Error Analysis of Motion Parameter Estimation from Image Sequences," Proc. Intl. Conf. on Computer Vision ICCV'87, pp. 703–707 (1987).

78. Yasumoto Y., and Medioni G., "Experiments in Estimation of 3–D Motion Parameters from a Sequence of Image Frames," Proc. IEEE Conf. Computer Vision and Pattern Recognition CVPR'85, pp. 89–94 (1985).

INDEX